GAS WELL DELIQUIFICATION

GAS WELL DELIQUIFICATION

SOLUTION TO GAS WELL LIQUID LOADING PROBLEMS

JAMES LEA
HENRY NICKENS
MICHAEL WELLS

Gulf Professional Publishing
an imprint of Elsevier Science

Gulf Professional Publishing is an imprint of Elsevier.

International Standard Book Number 0-7506-7724-4

British Library Cataloguing-in-Publication Data
A catalogue record for this book is available from the British Library.

The publisher offers special discounts on bulk orders of this book.

C.H.I.P.S.
10777 Mazoch Road
Weimar, TX 78962 U.S.A.
Tel: 979 263-5683
Fax: 979 263-5685
www.chipsbooks.com

For information on all Gulf Professional Publishing publications available, contact our World Wide Web home page at: http://www.gulfpp.com

10 9 8 7 6 5 4 3 2 1

Printed in the United States of America

CONTENTS

PREFACE

Most gas well streams contain water or condensate. As the well pressure and production rate decline, liquids begin to accumulate in the tubing or flow path. *Gas Well Deliquification* contains methods of predicting and analyzing this situation. Also presented are many proven methods that are used in the oil and gas industry to eliminate or reduce the effects of liquid loading so that gas well production can proceed with minimal interference. This collection of information should be helpful to many gas well producers.

James Lea
Henry Nickens
Michael Wells

CHAPTER 1

INTRODUCTION

1.1 INTRODUCTION

Liquid loading of a gas well is the inability of the produced gas to remove the produced liquids from the wellbore. Under this condition, produced liquids will accumulate in the wellbore leading to reduced production and shortening of the time until when the well will no longer produce.

This book deals with the recognition and operation of gas wells experiencing liquid loading. It presents materials on methods and tools to enable you to diagnose liquid loading problems and indicates how to operate your well more efficiently by reducing the detrimental effects of liquid loading on gas production.

This book will serve as a primer to introduce most of the possible and most frequently used methods that can help produce gas wells when liquids begin to become a problem. Liquid loading can be a problem in both high rate and low rate wells, depending on the tubular sizes, the surface pressure, and the amount of liquids being produced with the gas. In this book you will learn:

- How to recognize liquid loading when it occurs
- How to model gas well liquid loading
- How to design your well to minimize liquid loading effects
- What tools are available to assist you in the design and analysis of gas wells for liquid loading problems
- The best methods of minimizing the effects of liquids in lower velocity gas wells and the advantages and disadvantages of these methods
- How and why to apply various artificial lift methods for liquid removal

- What should be considered when selecting a lift method for liquid removal

1.2 MULTIPHASE FLOW IN A GAS WELL

To understand the effects of liquids in a gas well, we must understand how the liquid and gas phases interact under flowing conditions.

Multiphase flow in a vertical conduit is usually represented by four basic flow regimes as shown in Figure 1-1. A flow regime is determined by the velocity of the gas and liquid phases and the relative amounts of gas and liquid at any given point in the flowstream.

One or more of these regimes will be present at any given time in a well's history.

- **Bubble Flow**—The tubing is almost completely filled with liquid. Free gas is present as small bubbles, rising in the liquid. Liquid contacts the wall surface, and the bubbles serve only to reduce the density.

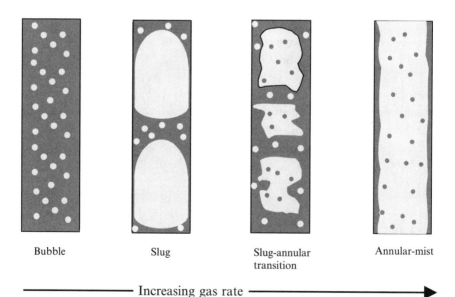

| Bubble | Slug | Slug-annular transition | Annular-mist |

———————— Increasing gas rate ————————►

Figure 1-1. Flow regimes in vertical multiphase flow.

- **Slug Flow**—Gas bubbles expand as they rise and coalesce into larger bubbles and then slugs. Liquid phase is still the continuous phase. The liquid film around the slugs may fall downward. Both gas and liquid significantly affect the pressure gradient.
- **Slug-Annular Transition**—The flow changes from continuous liquid to continuous gas phase. Some liquid may be entrained as droplets in the gas. Although gas dominates the pressure gradient, liquid effects are still significant.
- **Annular-Mist Flow**—Gas phase is continuous, and most of liquid is entrained in the gas as a mist. Although the pipe wall is coated with a thin film of liquid, the pressure gradient is determined predominately from the gas flow.

A gas well may go through any or all of these flow regimes during its lifetime. Figure 1-2 shows the progression of a typical gas well from initial production to end of life. In this illustration, it is assumed that the tubing end does not extend to the mid-perforations so that there is a section of casing from the tubing end to mid-perforations.

The well may initially have a high gas rate so that the flow regime is in mist flow in the tubing; however, it may be in bubble, transition, or slug flow below the tubing end to the mid-perforations. As time increases and

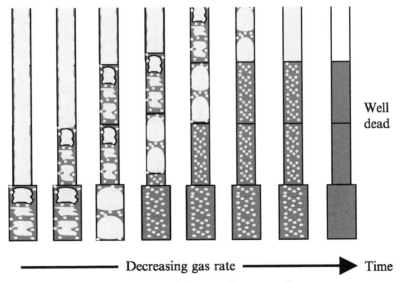

Decreasing gas rate ——————▶ Time

Figure 1-2. Life history of a gas well.

production declines, the flow regimes from the perforations to surface will change as the gas velocity decreases. Liquid production may also increase as the gas production declines. Flow at surface will remain in mist flow until the conditions change sufficiently at the surface so that the flow exhibits transition flow. At this point, the well production will become somewhat erratic, progressing to slug flow as the gas rate continues to decline. This transition will often be accompanied by a marked increase in the decline rate. The flow regime further downhole may be in bubble or slug flow, even though the surface production is in stable mist flow.

Eventually, the unstable slug flow at surface will transition to a stable, fairly steady production rate again as the gas rate declines further. This event occurs when the gas rate is too low to carry liquids to surface and simply bubbles up through a stagnant liquid column.

If corrective action is not taken, the well will continue to decline and will eventually log off. It is also possible that the well may continue to flow for a long period in a loaded condition and that gas produces up through liquids with no liquids rising to the surface.

1.3 WHAT IS LIQUID LOADING?

When gas flows to surface, the gas carries the liquids to the surface if the gas velocity is high enough. A high gas velocity results in a mist flow pattern in which the liquids are finely dispersed in the gas. This results in a low percentage by volume of liquids being present in the tubing (i.e., low liquid "holdup") or production conduit, resulting in a low pressure drop caused by the gravity component of the flowing fluids.

According to the Interstate Oil and Gas Compact Commission, in 2000, 411,793 stripper oil wells in the United States produced an average of 2.16 bpd and 223,707 stripper gas wells produced an average of 15.4 Mscf/D. For the lower-producing gas wells operating on the edge of profitability, optimization and reduction of liquid loading can mean the difference between production and shutting the well in. Liquid loading in gas wells is not limited, however, to the low rate producers; gas wells with large tubulars and/or high surface pressure can suffer from liquid loading even at high rates.

A well flowing at a high gas velocity can have a high pressure drop caused by friction; however, for higher gas rates, the pressure drop caused by accumulated liquids in the conduit is relatively low. This subject is discussed in greater detail later in the book.

As the velocity of the gas in the production conduit drops with time, the velocity of the liquids carried by the gas declines even faster. As a result, flow patterns of liquids on the walls of the conduit, liquid slugs forming in the conduit, and eventually liquids accumulating in the bottom of the well occur; all of which increase the percentage of liquids in the conduit while the well is flowing. The presence of more liquids accumulating in the production conduit while the well is flowing can either slow production or stop gas production altogether.

Few gas wells produce completely dry gas. Under some conditions, gas wells will produce liquids directly into the wellbore. Both hydrocarbons (condensate) and water may condense from the gas stream as the temperature and pressure change during travel to the surface. In some cases, fluids may come into the wellbore as a result of coning water from an underlying zone or from other sources.

Most of the methods used to remove liquids from gas wells do not depend on the source of the liquids. However, if a remediation method is considered that addresses condensation only, then it must be determined that this is indeed the source of the liquid loading. If not, the remediation will be unsuccessful.

1.4 PROBLEMS CAUSED BY LIQUID LOADING

Liquid loading can lead to erratic, slugging flow and to decreased production from the well. The well may eventually die if the liquids are not removed continuously, or the well may produce at a lower rate than possible.

If the gas rate is high enough to continually produce most or all of the liquids, the wellbore formation pressure and production rate will reach a stable equilibrium operating point. The well will produce at a rate that can be predicted by the reservoir inflow performance relationship (IPR) curve (see Chapter 4).

If the gas rate is too low, the tubing pressure gradient becomes larger because of the liquid accumulation resulting in increased pressure on the formation. As the backpressure on the formation increases, the produced rate from the reservoir decreases and may drop below the so-called "gas critical rate" required to continuously remove the liquid. More liquids will accumulate in the wellbore, and the increased bottomhole pressure will reduce production or may kill the well.

Late in the life of a well, liquid may stand over the perforations with the gas bubbling through the liquid to the surface. The gas is producing at a low but steady rate, and no liquids may be coming to the surface. If this was observed without any knowledge of past well history, one might assume that the well is only a low gas producer, not liquid loaded.

All gas wells that produce liquids—whether in high or low permeability formations—will eventually experience liquid loading with reservoir depletion. Even wells with very high gas-liquid ratios (GLR) and small liquid rates can load up if the gas velocity is low. This condition is typical of very tight formation (low permeability) gas wells that produce at low gas rates and have low gas velocities in the tubing. Some wells may be completed and produce a considerable gas rate through large tubulars, but may be liquid loaded from the first day of production. Lea and Tighe[1] and Libson and Henry[2] provide an introduction to loading and some discussion of field problems and solutions.

1.5 DELIQUEFYING TECHNIQUES

The following list[3] (modified) introduces some of the possible methods to deliquefy gas wells that are discussed here. These methods may be used singly or in combination. This list is based roughly on the static reservoir pressure.

Each of these methods is discussed in some detail. This list is not presented as being 100% complete. Special methods, such as using a pumping system to inject water below a packer to allow gas to flow up the casing-tubing annulus, are covered in the chapters on de-watering using beam and ESP pumping systems. Depth considerations and certain economic considerations also are not detailed.

The method that is most economic for the longest period of operation is the optimum method. The criteria for selecting the optimum method are: methods in similar fields that are used successfully, vendor equipment availability, reliability of equipment, manpower required to operate the equipment, and lifting capacity.

- Reservoir Pressure >1500 psi
  - Evaluate best natural flow of the well
  - Use Nodal Analysis to evaluate the tubing size for friction and future loading effects
  - Consider possible coiled tubing use

- Evaluate surface tubing pressure and seek low values for maximum production
- Consider annular flow or annular and tubing flow to reduce friction effects
• Reservoir Pressure between 500 and 1500 psi
  - These medium pressure wells may still flow using relatively smaller conduits and low surface pressures to keep flow velocities above a "critical" rate.
  - Low pressure systems
  - Plunger lift
  - Small tubing
  - Reduce surface pressure
  - Regular swabbing for short flow periods
  - Pit blow-downs (environmentally unacceptable)
  - Surfactant soap sticks down the tubing or liquids injected down tubing or casing
  - Reservoir flooding to boost pressures
• Reservoir Pressure between 500 and 1500 psi
  - Lower pressure systems
  - Plunger lift—can operate with large tubing
  - Small tubing
  - Reduce surface pressure
  - Surfactants
  - Siphon strings; usually smaller diameter
  - Rod pumps on pump-off control
  - Intermittent gas lift
  - Hydraulic jet or reciprocating hydraulic pump
  - Swabbing
  - Reservoir flooding
• Very Low Pressure Systems (Reservoir Pressure <150 psi)
  - Rod pumps
  - Plunger in some cases
  - Siphon strings
  - Reduce surface pressure
  - Intermittent gas lift; chamber lift
  - Hydraulic jet or reciprocating hydraulic pump
  - Swabbing
  - Surfactants
  - Reservoir flooding

1.6 SOURCE OF LIQUIDS IN A PRODUCING GAS WELL

Many gas wells produce not only gas but also condensate and water. If the reservoir pressure has decreased below the dew point, the condensate is produced with the gas as a liquid; if the reservoir pressure is above the dew point, the condensate enters the wellbore in the vapor phase with the gas and condenses as a liquid in the tubing or separator.

Produced water may have several sources.

- Water may be coned in from an aqueous zone above or below the producing zone.
- If the reservoir has aquifer support, the encroaching water will eventually reach the wellbore.
- Water may enter the wellbore from another producing zone, which could be separated some distance from the gas zone.
- Free formation water may be produced with the gas.
- Water and/or hydrocarbons may enter the wellbore in the vapor phase with the gas and condense out as a liquid in the tubing.

1.6.1 Water Coning

If the gas rate is high enough, then the gas may entrain water production from an underlying zone, even if the well is not perforated in the water zone. A horizontal well greatly reduces gradients between the gas zone and an underlying water zone; however, the same phenomenon can occur at very high rates, although it is usually termed "cresting" instead of "coning."

1.6.2 Aquifer Water

Pressure support from an aquifer will eventually allow water production to reach the wellbore, giving rise to liquid-loading problems.

1.6.3 Water Produced from Another Zone

Another zone may produce into the wellbore with an open hole or in a well with several sections perforated. The reverse situation that takes advantage of this situation is to have a water zone below the gas zone and by using pumps or gravity, inject water into an underlying zone and allow gas to flow to the surface with no loading problems.

1.6.4 Free Formation Water

From whatever the source, it is possible that water comes in the perforations with the gas. This situation can be caused by various thin layers of gas and liquids or for other reasons.

1.6.5 Water of Condensation

If saturated or partially saturated gas enters the well, the perforations have no liquids entering, but condensation can occur higher in the well. This situation can cause a high gradient in the flowstring where the condensation occurs and also, depending on velocities, liquids can fall back and accumulate over the perforations or pay zone.

Everyone has experienced the phenomenon of water condensing from the atmosphere (i.e., rain). At any given pressure and temperature, a certain amount of water vapor will be in equilibrium with the atmospheric gases. As temperature decreases or pressure increases, the amount of equilibrium water vapor decreases, and any excess water vapor will condense to the liquid phase to maintain equilibrium. If temperature increases or pressure decreases, free liquid water (if present) will evaporate to the vapor phase to maintain equilibrium.

A similar phenomenon occurs in hydrocarbon gas. For a given reservoir pressure and temperature, the produced gas may contain a certain amount of water vapor. Figure 1-3 shows an example of the solubility of water in natural gas in STB/MMscf. Note the rapid increase in water content as reservoir pressure declines below 500 psi.

The water will remain in the vapor phase until temperature and pressure conditions drop below the dew point. When this occurs, some of the water vapor will condense to the liquid phase. If the condensation occurs in the wellbore and if the gas velocity is below the critical rate required to carry the liquid water, then liquids will accumulate in the wellbore, and liquid loading will occur.

1.6.6 Hydrocarbon Condensates

Hydrocarbons can also enter the well with the gas with the production in the vapor stage. If the reservoir temperature is above the cricondentherm, then no liquids will be in the reservoir, but they can drop out in the wellbore just as water condensation can occur.

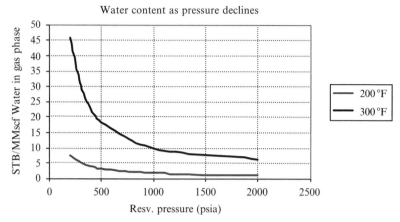

Figure 1-3. Water solubility in natural gas.

Table 1-1
Water Solubility in Natural Gas

Location	P/T	Water Content (STB/MMscf)	Water Condensed in Tubing (STB/MMscf)
Surface	150 psi/100 °F	0.86	—
Reservoir	3500 psi/200 °F	0.73	0
Reservoir	1000 psi/200 °F	1.75	0.89
Reservoir	750 psi/200 °F	2.22	1.36
Reservoir	500 psi/200 °F	3.17	2.31
Reservoir	250 psi/200 °F	6.07	5.21

Even if the gas velocity is sufficient to remove the condensed water, corrosion problems may occur at the point in the wellbore where condensation first occurs. Condensed water can be identified because it should have a much lower or no salt content compared to reservoir water. Normally, we would assume pure water in the vapor phase before condensation.

Example 1.1: Water Solubility in Natural Gas

Consider the following typical example for a gas well producing initially from a reservoir at 3500 psi and 200 °F producing at wellhead conditions of 150 psi and 100 °F. In this example, we assume that the wellhead conditions remain constant as shown above in Table 1-1.

As the reservoir pressure declines, the water condensing out in the tubing increases. Because the gas rate will decline as the reservoir pressure decreases, we have the situation of decreasing gas rate coupled with increasing liquid production—liquid loading will inevitably occur.

REFERENCES

1. Lea, J. F., and Tighe, R. E., "Gas Well Operation with Liquid Production," SPE 11583, presented at the 1983 Production Operation Symposium, Oklahoma City, OK, February 27–March 1, 1983.

2. Libson, T. N., and Henry, J. T., "Case Histories: Identification of and Remedial Action for Liquid Loading in Gas Wells-Intermediate Shelf Gas Play," *Journal of Petroleum Technology*, April 1980, pp 685–693.

3. Coleman, S. B., et al., "A New Look at Predicting Gas Well Liquid Load-Up," *Journal of Petroleum Technology*, March 1991, pp 329–332.

RECOGNIZING SYMPTOMS OF LIQUID LOADING IN GAS WELLS

2.1 INTRODUCTION

Over the life of a gas well it is likely that the volume of liquids being produced will increase while the volume of gas being produced will drop off. These situations usually result in the accumulation of liquids in the wellbore until eventually the well dies or flows erratically at a much lower rate. If diagnosed early, losses in gas production can be minimized by implementing one of the many methods available to artificially lift the liquids from the well.

On the other hand, if liquid loading in the wellbore goes unnoticed, the liquids could accumulate in the wellbore and the adjoining reservoir, possibly causing temporary or even permanent damage. It is vital, therefore, that the effects caused by liquid loading are detected early to prevent loss of production and possible reservoir damage.

This chapter will discuss the symptoms that indicate when a gas well is having problems with liquid loading. Emphasis is placed on symptoms that are typically available for inspection in the field. Although some of these are more obvious than others, all lend themselves to more exacting methods of well analysis described in the following chapters.

These symptoms indicate a well is liquid loading:

- Presence of recorded through the gas measuring device pressure spikes
- Erratic production and increase in decline rate
- Tubing pressure decreases as casing pressure increases
- Pressure survey shows a sharp, distinct change in pressure gradient
- Annular heading
- Liquid production ceases

2.2 PRESENCE OF ORIFICE PRESSURE SPIKES

One of the most common methods available to detect liquid loading is recorded production data from an automated data collection system or two pen pressure recorder. These devices record the measurement of gas flow rate through an orifice over time. Typically, when a well produces liquids without loading problems, the liquids are produced in the gas stream as small droplets (mist flow) and have little effect on the orifice pressure drop. When a liquid slug passes through the orifice, however, the relatively high density of the liquid causes a pressure spike. A pressure spike on a plot of orifice pressure drop usually indicates that liquids are beginning to accumulate in the wellbore and/or the flowline and are being produced erratically as some of the liquids reach the surface as slugs.

This phenomenon is depicted in Figure 2-1 on a two pen recorder showing a well that is producing liquids normally in mist flow on the left and a well beginning to experience liquid loading problems by producing the liquids in slugs on the right.

When liquids begin to accumulate in the wellbore, the pressure spikes on the recorder become more frequent. Eventually, the surface tubing pressure starts to decrease because the liquid head holds back the reservoir pressure. In addition, the gas flow begins to decline at a rate faster than the prior production decline rate. This rapid drop in production and drop in surface tubing pressure, accompanied by the ragged two pen recorder charts, is a sure indication of the onset of liquid loading. The well chart shown in Figure 2-2 indicates severe liquid loading. Figure 2-3 shows a well where the liquid loading situation has been improved but not completely solved as noted by the smaller more consistent spikes.

2.3 DECLINE CURVE ANALYSIS

The shape of a well's decline curve can indicate downhole liquid loading problems. Decline curves should be analyzed over time, looking for changes in the general trend. Figure 2-4 shows two decline curves. The smooth exponential type decline curve is characteristic of normal gas-only production considering reservoir depletion. The sharply fluctuating curve is indicative of liquid loading in the wellbore and, in this case, shows that the well will deplete much earlier than reservoir considerations alone would indicate. When decline curve trends are analyzed

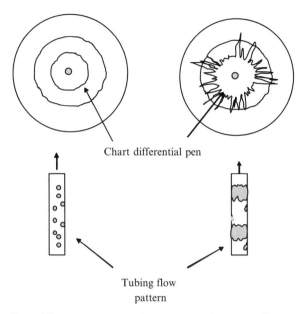

Chart differential pen

Tubing flow
pattern

Figure 2-1. Effect of flow regime on orifice pressure drop. Mist flow (L) vs. slug flow (R) in tubing.

for long periods, wells experiencing liquid-loading problems will typically show a sudden departure from the existing curve to a new, steeper slope. The new curve will indicate well abandonment far earlier than the original curve, providing a method to determine the extent of lost reserves as a result of liquid loading. The remedial lift methods described in this book can help restore production to the original decline curve slope.

2.4 DROP IN TUBING PRESSURE WITH RISE IN CASING PRESSURE

If liquids begin to accumulate in the bottom of the wellbore, the added pressure head on the formation lowers the surface tubing pressure. In addition, as the liquid production increases, the added liquid in the tubing being carried by the gas (liquid hold-up) increases the gradient in the tubing and again provides more backpressure against the formation, thereby reducing the surface tubing pressure.

In packerless completions where this phenomenon can be observed, the presence of liquids in the tubing is shown as an increase in the surface casing pressure as the fluids bring the reservoir to a lower flow, higher

FIELD GAS CHART

LOADED WELL

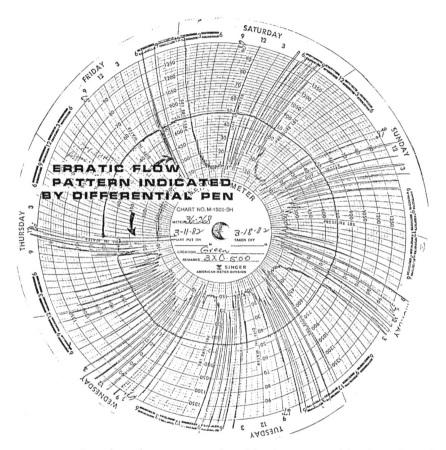

Figure 2-2. Gas chart showing severe liquid loading as noted by the indicated slugs of liquid.

pressure production point. As gas is produced from the reservoir, gas percolates into the tubing casing annulus. This gas is exposed to the higher formation pressure, causing an increase in the surface casing pressure. Therefore, a decrease in tubing pressure and a corresponding increase in casing pressure are indicators of liquid loading. Although these effects are illustrated in Figure 2-5, the changes may not be linear with time as shown in this illustration.

Estimates of the tubing pressure gradient can be made in a flowing well without a packer by measuring the difference in the tubing and casing

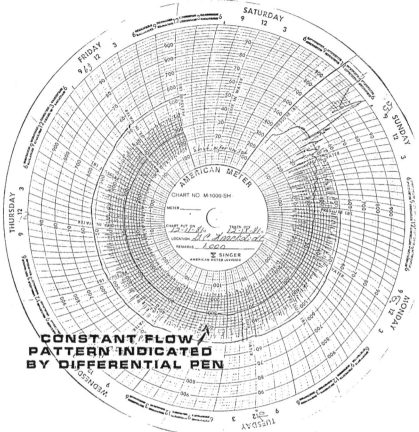

Figure 2-3. Gas chart showing much less indication of liquid loading.

pressures. In a packerless production well, the free gas will separate from the liquids in the wellbore and will rise into the annulus. The fluid level in a flowing well will remain at the tubing intake except when "heading" occurs or if a tubing leak is present.

During "heading," the liquid level in the annulus periodically rises above and then falls back to the tubing intake. In a flowing well, however, the difference in the surface casing and tubing pressures is an indication of the pressure loss in the production tubing. The weight of the gas column in the casing can be easily computed (see Appendix C). Comparing the casing and tubing pressure difference with a dry gas

Decline curve as indicator of liquid loading rate

Rate

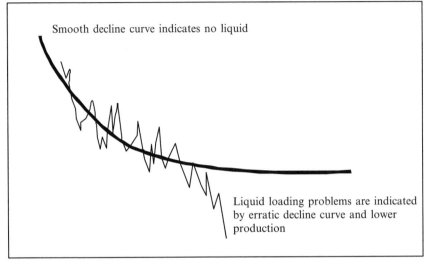

Time

Figure 2-4. Decline curve analysis.

gradient for the well can give an estimate of the higher tubing gradient caused by liquids accumulating or by loading the tubing.

2.5 PRESSURE SURVEY SHOWING TUBING LIQUID LEVEL

Flowing or static well pressure surveys are perhaps the most accurate method available to determine the liquid level in a gas well and thereby whether the well is loading with liquids. Pressure surveys measure the pressure with depth of the well either while shut-in or while flowing. The measured pressure gradient is a direct function of the density of the medium and the depth; and, for a single static fluid, the pressure with depth should be nearly linear.

Because the density of the gas is significantly lower than the density of water or condensate, the measured gradient curve will exhibit a sharp change of slope when the tool encounters standing liquid in the tubing. Thus, the pressure survey provides an accurate means of determining the liquid level in the wellbore.

Figure 2-6 illustrates the basic principle associated with the pressure survey. The gas and liquid production rates and accumulations can

Liquid accumulation in tubing

Pressure

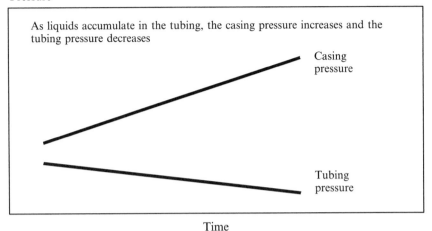

Time

Figure 2-5. Casing and tubing pressure indicators.

change the slopes measured by the survey, giving a higher gas gradient due to the presence of some liquids dispersed and a lower liquid gradient due to the presence of gas in the liquid. The liquid level in a shut-in gas well can also be measured acoustically by shooting a liquid level down the tubing.

The fluid in the tubing in a well that produces both liquids and gases exhibits a complicated two-phase flow regime that depends on the flow rate and the amount of each constituent phase present. The flowing pressure survey data obtained in two-phase flow regime are not necessarily linear as indicated above. When the measured pressure gradient is not linear but shows a continuously increasing pressure with depth, pressure gradient data alone are not sufficient to determine if liquid loading is becoming a problem.

In these cases, it may be necessary to either repeat the pressure survey at other conditions or to use techniques described later in this text to compute the gradient in smaller tubing sizes or lower surface pressures to determine if liquids are accumulating. Often, the pressure deflection brought about by standing liquid in the tubing can be masked by higher flow rates in small tubing. The added frictional pressure loss in these cases can "mask" the inflection point caused by the liquid interface. Large tubing usually means a lower frictional pressure loss (depends on the flow rate) and, as a result, typically produces a sharp deflection in the pressure survey curve. Some wells have a tapered tubing string. In this case, a change in tubing cross-sectional flow area may cause a change

Pressure survey to determine liquid loading

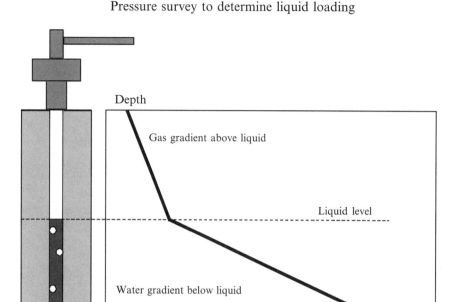

Figure 2-6. Pressure survey schematic.

in flow regime at the point where the flow area changes, with a resultant change in the pressure gradient. This situation may appear in a gradient survey as a change in the slope of the pressure-depth plot at the depth of the tubing area change and should not be confused with the gas-liquid interface at the depth of the liquid level.

An estimate of the volume of liquid production can be made by comparing the tubing pressure loss in a well producing liquids with one producing only or near dry gas. In a flowing well, the bottom-hole pressure (BHP) is equal to the pressure drop in the tubing (or annulus if flowing up the annulus), plus the wellhead pressure. The presence of the liquid in the production stream always increases the tubing pressure gradient. At low gas rates, the proportional increase of pressure loss in the tubing caused by liquids is higher than at high gas rates. The variance then allows one, with a productivity expression for gas flow from the reservoir, to see how more production is possible if the pressure increase caused by liquid loading is mitigated. Chapter 4 provides illustrations of the tubing performance curve for gas with some liquids intersecting a reservoir inflow curve as a method of predicting gas well production.

2.6 WELL PERFORMANCE MONITORING

A method is presented for displaying the minimum lift (and erosional gas rate) directly on the wellhead backpressure curve.[1] These curves help identify when liquid loading (or erosional rates) threatens to reduce production. An overlay technique is identified whereby a minimum lift "type-curve" is generated for an entire field or for a particular set of operating conditions.

2.7 ANNULUS HEADING

Some gas wells without packers establish low-frequency pressure oscillations, which can persist over several hours or days. These oscillations are indicative of the accumulation of produced liquids in the wellbore and have reportedly curtailed production by more than 40%. Figure 2-7 illustrates this possible oscillatory behavior for a typical packerless gas well.

The process is described by W. E. Gilbert.[2] It is essentially a low flow rate process with a high annulus level. Later a quick high gas flow rate occurs with a low annulus liquid level, which temporarily exhausts gas rapidly and wastes some of the reservoir flowing pressure because liquids are not carried with the burst of gas. Although not strictly liquid loading, such as an increased concentration of liquid in the tubing or flow path, the oscillations of the liquid level and gas pressure in the annulus can significantly reduce production if they continue unchecked.

2.7.1 Heading Cycle without Packer

The steps of the annulus heading cycle shown in Figure 2-7 are listed below. The cycle description begins with the annulus fluid level at the peak height.

1. Gas trickles into the annulus and slowly displaces annulus liquid into the tubing, lowering the annulus liquid level and decreasing casing pressure.
2. The well is still producing at a low rate because the tubing column is "heavy" because of the diversion of some production gas into the

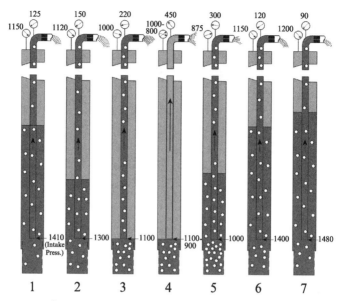

Figure 2-7. Low frequency pressure oscillations in a gas well producing liquids.

casing and because of the added production of casing liquids in the tubing caused by the accumulation of annulus gas pressure. Annulus pressure is still decreasing as more annulus fluid is displaced into the tubing.

3. The pressure in the annulus continues to drop. The annulus liquid level drops to the tubing inlet as the liquid is produced out of the annulus. Gas flows into the tubing. The weight of the tubing column is reduced, because the gas from the formation is now produced up the tubing and is no longer trickling into the annulus. Liquids from the annulus are no longer being diverted into the production stream.

4. The tubing gradient drops still further because of produced gas in the flow stream, lowering the bottom hole tubing pressure and allowing dry gas from the annulus to "blow around" into the tubing. The production from the reservoir is also increased, depleting the reservoir near the wellbore much more than the other times in this cycle. For a short period, the well produces at a higher than normal rate but with relatively small amounts of liquid. Because the liquid production is low or nonexistent during the high flow rate period, the energy provided by the high gas rate is wasted in the sense that gas flows without lifting some liquids.

5. The reservoir again starts to produce liquids, and the gas production drops. The gas stored in the annulus is depleted, and the tubing and casing annulus begin to load with liquids. As the liquid level rises in the annulus, gas also begins to percolate into the annulus. Because gas is now diverted into the annulus, the gradient in the tubing increases, adding extra force against the reservoir and lowering the production rate.

6. Liquid still flows into the tubing at a higher rate than can be carried out of the tubing by the gas flow. Liquids continue to accumulate in the bottom of the well. Some gas is migrating into the casing and tubing annulus.

7. The rate of production of liquids at the surface is in balance with the rate of liquid production at the formation. Gas continues to migrate into the annulus until the annular pressure peaks, again forcing liquid into the tubing and repeating the cycle.

Although this is not "liquid loading" in the usual sense, it is caused by an instability in the casing and tubing annulus pressures, which leads to ups and downs in the production. The production of formation liquids under the cyclic behavior described above is inefficient because a portion of the cycle produces a high volume of the gas with very little lifting of the liquids. In some instances, it is possible to choke back the well to control the cyclic behavior; however, this method also cuts back production by increasing the average bottomhole pressure.

2.7.2 Heading Cycle Controls

Methods used to control the oscillations while maintaining a high average flow rate are to install a downhole packer to prevent the gas from migrating into the annulus or a surface controller that monitors the pressures, preventing casing pressure buildup.

Use of a surface controller to counter the natural oscillations brought about by gas accumulation in the annulus is illustrated in Figure 2-8.

1. At the start of the flowing period, the tubing is opened by the rising casing pressure, which actuates the motor valve. The column of gas collected in the upper part of the tubing is produced, and the subsequent reduction of pressure ensures flow of the fluid mixture in the tubing below the gas column.

2. The tubing pressure trends downward while fluid is being displaced out of the annulus.

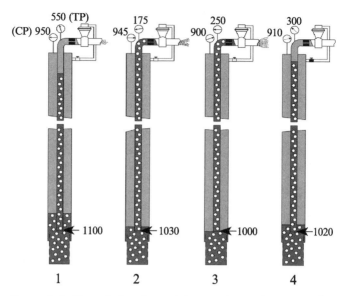

Figure 2-8. Control of well instabilities with a surface controller.

3. The tubing pressure then rises as annulus gas starts to break around the foot of the tubing.
4. When the casing pressure reaches the predetermined minimum, the motor valve closes the tubing outlet (or pinches it back), but flow into the well from the reservoir continues with very little decrease in rate. This production includes both gas and liquid, which flow into the annulus, effectively filling it. The tubing pressure continues to rise. The casing pressure, which is directly related to the amount of gas stored in the annulus, also increases in response to gas and liquid entering the well. When the casing pressure reaches the predetermined maximum, the cycle is repeated by opening the motor valve or opening it to a wider flow area for the gas.

By smoothing out the flow, surface controllers can be used to increase the rate of flow and extend the flowing life of wells that have reached the annulus heading stage. Although this type of control produces formation liquids less effectively than pumping systems, it is a good option when other lifting methods are not feasible. The use of a surface controller on heading wells can increase production and can prolong the life of wells not equipped with downhole packers without expensive workovers. These controllers are applicable for wells without packers.

2.8 LIQUID PRODUCTION CEASES

Some high-rate gas wells readily produce liquids for a time and then drop off to much lower rates. As the gas production declines, the liquid production can cease. In these cases, the well is producing gas at rates below the "critical" rate that can transport the liquids to the surface. The result is that the liquids continue to accumulate in the wellbore, and the gas bubbles through the accumulated liquids. Depending on the accumulation of liquids and the well pressure, the well can either cease to flow or the gas can bubble up through the liquids. However, the gas rate has dropped to a value where liquids are no longer transported up the tubing.

The best method to analyze this type of low flow well response is to calculate a minimum critical velocity in the tubing or the minimum gas velocity required to carry liquids to the surface. If the flow is well below what is necessary to lift liquids, and especially if the flow rate is low in large-diameter tubing, then the possibility of gas bubbling through accumulated liquids should be investigated. Pumping the liquids out of the well, recompleting the well with smaller tubing, or using coil tubing to inject N_2 may be solutions for this low flow rate situation.

Wireline pressure surveys can also be used to determine if there is standing liquid in the wellbore. These methods will be discussed in later chapters. It is also possible to shoot an acoustically measured fluid level down the tubing if the flow does not interfere with the acoustical signals received from reflections of a pressure pulse at the surface or if the fluid shot is done quickly and periodically after shut-in of the well.

2.9 SUMMARY

Several symptoms of wells suffering from liquid loading have been illustrated. These indicators provide early warning of liquid-loading problems that can hamper production and sometimes can permanently damage the reservoir. These indicators should be monitored on a regular basis to prevent loss of production. Methods to analytically predict loading problems and the subsequent remedial action will be discussed in later chapters of this book.

REFERENCES

1. Thrasher, T. S., "Well Performance Monitoring: Case Histories," SPE 26181, presented at the SPE Gas Technology Symposium, Calgary, Alberta, Canada, June 28–30, 1993.

2. Gilbert, W. E., "Flowing and Gas-Lift Well Performance," presented at the spring meeting of the Pacific Coast District, Division of Production, Los Angeles, May 1954, Drilling and Production Practice, pp. 126–157.

CHAPTER 3

CRITICAL VELOCITY

3.1 INTRODUCTION

To effectively plan and design for gas well liquid-loading problems, it is essential to be able to predict accurately when a particular well might begin to experience excessive liquid loading. In the next chapter, Nodal Analysis (™ of Macco-Schlumberger) techniques are presented that can be used to predict when liquid loading problems and well flow stability occur. This chapter presents the relatively simple "critical velocity" concept used to predict the onset of liquid loading. This technique was developed from a substantial accumulation of well data and has been shown to be reasonably accurate for vertical wells. The method of calculating a critical velocity will be shown to be applicable at any point in the well. It should be used in conjunction with methods of Nodal Analysis if possible.

3.2 CRITICAL FLOW CONCEPTS

The transport of liquids in near vertical wells is governed primarily by two complementing physical processes before liquid loading becomes more predominate. As the well becomes weaker, other flow regimes, such as slug flow and then bubble flow, begin to appear.

3.2.1 Turner Droplet Model

It is generally believed that liquids are lifted in the gas flow velocity regimes as individual particles and transported as a liquid film along the tubing wall by the shear stress at the interface between the gas and

Liquid transport in a vertical gas well

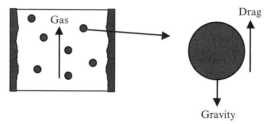

Figure 3-1. Illustrations of concepts investigated for defining "critical velocity."

the liquid before the onset of severe liquid loading. These mechanisms were investigated by Turner et al.,[1] who evaluated two correlations developed on the basis of the two transport mechanisms using a large experimental database. Turner discovered that liquid loading could be predicted by a droplet model that showed when droplets move up (gas flow above critical velocity) or down (gas flow below critical velocity).

Turner et al. developed a simple correlation to predict the so-called "critical velocity" in near vertical gas wells assuming the droplet model. In this model, the droplet weight acts downward, and the drag force from the gas acts upward (Figure 3-1). When the drag is equal to the weight, the gas velocity is at *"critical."* Theoretically, at the critical velocity, the droplet would be suspended in the gas stream, moving neither upward nor downward. Below the critical velocity, the droplet falls, and liquids accumulate in the wellbore.

In practice, the critical velocity is generally defined as the minimum gas velocity in the production tubing required to move droplets upward. A "velocity string" is often used to reduce the tubing size until the critical velocity is obtained. Lowering the surface pressure (e.g., by compression) also increases the actual gas velocity.

Turner's correlation was tested against a large number of real well data having surface flowing pressures mostly higher than 1000 psi. Considering that Turner made his correlation using high surface well data, it is not clear how accurate his correlations would be at lower wellhead pressures. For surface pressures below about 1000 psi, the use of Turner's correlations may become more suspect.

Two variations of the correlation were developed: one for the transport of water and the other for condensate. The fundamental equations derived by Turner et al. were found to underpredict the critical velocity from the database of well data. To better match the collection of measured field data, Turner et al. adjusted the theoretical equations for required velocity upward by 20%. From Turner et al.'s[1] original paper, after the 20% empirical adjustment, the critical velocity for condensate and water was presented as

$$v_{g,condensate} = \frac{4.02(45 - 0.0031_P)^{1/4}}{(0.0031_P)^{1/2}} \; \text{ft/sec} \qquad (3\text{-}1)$$

$$v_{g,water} = \frac{5.62(67 - 0.0031_P)^{1/4}}{(0.0031_P)^{1/2}} \; \text{ft/sec} \qquad (3\text{-}2)$$

where $P = \text{psi}$

The theoretical equation from Turner et al. for critical velocity V_t to lift a liquid (see Appendix A) is

$$V_t = \frac{1.593\sigma^{1/4}(\rho_1 - \rho_g)^{1/4}}{\rho_g^{1/2}} \; \text{ft/sec} \qquad (3\text{-}3)$$

where $\sigma = $ surface tension, dynes/cm $\rho = $ density, lbm/ft^3

Inserting typical values of:

Surface Tension   20 and 60 dyne/cm for condensate and water, respectively
Density   45 and 67 lbm/ft^3 for condensate and water, respectively
Gas Z factor   0.9

$$
\begin{aligned}
V_{t,condensate} &= \frac{1.593(20)^{1/4}(45 - .00279P/Z)^{1/4}}{(.00279P/Z)^{1/2}} \\
&= \frac{3.368(45 - .00279P/Z)^{1/4}}{(.00279P/Z)^{1/2}}
\end{aligned}
$$

$$
\begin{aligned}
V_{t,water} &= \frac{1.593(60)^{1/4}(67 - .00279P/Z)^{1/4}}{(.00279P/Z)^{1/2}} \\
&= \frac{4.43(67 - .00279P/Z)^{1/4}}{(.00279P/Z)^{1/2}}
\end{aligned}
$$

Inserting $Z = 0.9$ and multiplying by 1.2 to adjust the correlation to fit Turner et al.'s data gives:

$$V_{t,condensate} = \frac{4.043(45 - .0031P)^{1/4}}{(.0031P)^{1/2}}$$

$$V_{t,water} = \frac{5.321(67 - .0031P)^{1/4}}{(.0031P)^{1/2}}$$

Turner et al. give 4.02 and 5.62 in their paper for these equations.

These equations predict the minimum critical velocity required to transport liquids in a vertical wellbore. They are used most often and most easily at the wellhead, with P being the flowing wellhead pressure. When both water and condensate are produced by the well, Turner et al. recommend using the correlation developed for water, because water is heavier and requires a higher critical velocity. Gas wells having production velocities below that predicted by the above equations would then be less than required to prevent the well from loading with liquids. The actual rate of liquids produced does not appear in this correlation, and the predicted terminal velocity is not a function of the rate of liquid production.

3.2.2 Critical Rate

Although critical velocity is the controlling factor, one usually thinks of gas wells in terms of production rate in multiples of scf/D rather than velocity in the wellbore. These equations are easily converted into a more useful form by computing a critical well flow rate. From the critical velocity V_g, the critical gas flow rate q_g may be computed from:

$$q_g \text{ (MMscf/D)} = \frac{3.067 P V_g A}{(T + 460)Z} \tag{3-4}$$

where $A = \frac{(\pi)d_{ti}^2}{4 \times 144}$ ft^2
   T = surface temperature, °F
   P = surface pressure, psi
   A = tubing cross-sectional area
   d_{ti} = tubing ID, inches

Introducing these values into Turner et al.'s equations gives the following:

$$q_{t,condensate}(\text{MMscf}/\text{D}) = \frac{.0676 P d_{ti}^2}{(T+460)Z} \frac{(45 - .0031P)^{1/4}}{(.0031P)^{1/2}}$$

$$q_{t,water}(\text{MMscf}/\text{D}) = \frac{.0890 P d_{ti}^2}{(T+460)Z} \frac{(67 - .0031P)^{1/4}}{(.0031P)^{1/2}}$$

These equations can be used to compute the critical gas flow rate required to transport either water or condensate. Again, when both liquid phases are present, the water correlation is recommended. If the actual flow rate of the well is greater than the critical rate computed by the above equation then liquid loading would not be expected.

3.2.3 Critical Tubing Diameter

It is also useful to rearrange the above expressions, solving for the maximum tubing diameter that a well of a given flow rate can withstand without loading with liquids. This maximum tubing is termed the "critical tubing diameter" and corresponds to the minimum critical velocity. The critical tubing diameter for water or condensate is shown below where the critical velocity of gas, V_g, is for either condensate or water.

$$d_{ti,}inches = \sqrt{\frac{59.94 q_g (T+460)Z}{P V_g}}$$

3.2.4 Critical Rate for Low Pressure Wells—Coleman Model

Recall that the above relations were developed from data for surface tubing pressures mostly greater than 1000 psi. For lower surface tubing pressures, Coleman et al.[2] have developed similar relationships for the minimum critical flow rate for both water and liquid. In essence, the formulas from Coleman et al. (to fit their new lower wellhead pressure data—typically less than 1000 psi) are identical to Turner et al.'s equations but without the 1.2 adjustment to fit Turner's data. With the same data defaults given to develop Turner et al.'s equations, the Coleman et al.[2] equations for minimum critical velocity and flow rate are:

$$V_{t,condensate} = \frac{3.369(45 - .0031P)^{1/4}}{(.0031P)^{1/2}}$$

$$V_{t,water} = \frac{4.434(67 - .0031P)^{1/4}}{(.0031P)^{1/2}}$$

$$q_{t,condensate}(\text{MMscf}/\text{D}) = \frac{.0563Pd_{ti}^2}{(T + 460)Z} \frac{(45 - .0031P)^{1/4}}{(.0031P)^{1/2}}$$

$$q_{t,water}(\text{MMscf}/\text{D}) = \frac{.0742Pd_{ti}^2}{(T + 460)Z} \frac{(67 - .0031P)^{1/4}}{(.0031P)^{1/2}}$$

If the original equations of Turner were used, however, the coefficients would be 4.02 and 5.62 both divided by 1.2 to get the Coleman et al. equations, so there can be some confusion. The concern is that even if some slight errors in the Turner et al. development are present, the equations with the coefficients have been used in industry with success, and the question is "are the original coefficients better than if they are corrected or not?"

Example 3.1: Calculate the Critical Rate Using Turner et al. and Coleman et al.

Well surface pressure = 400 psia
Well surface flowing temperature = 120 °F
Water is the produced liquid
Water density = 67 lbm/ft³
Water surface tension = 60 dyne/cm
Gas gravity = 0.6

Gas compressibility factor for simplicity $= 0.9$
Production string $= 2\ 3/8$-inch tubing with 1.995 in ID, A $= .0217$ ft^2
Production $= .6$ MMscf/D

Critical Rate by Coleman et al.
   Calculate the gas density:

$$\rho_g = \frac{M_{air}\gamma_g P}{R(T+460)Z} = \frac{28.97\gamma_g P}{10.73(T+460)Z} = 2.7\frac{0.6 \times 400}{580 \times .9} = 1.24\ \text{lbm/ft}^3$$

$$V_g = \frac{1.593\sigma^{1/4}(\rho_1 - \rho_g)^{1/4}}{\rho_g^{1/2}} = \frac{1.593\ 60^{1/4}(67 - 1.24)^{1/4}}{1.24^{1/2}} = 11.30\ \text{ft/sec}$$

$$q_{t,water} = \frac{3.067P\ AV_g}{(T+460)Z} = \frac{3.067 \times 400 \times .0217 \times 11.30}{(120+460) \times 0.9} = .575\ \text{MMscf/D}$$

Critical Rate by Turner et al.
   Because the Turner and Coleman variations of the critical rate equation differ only in the 20% adjustment factor applied by Turner for his high pressure data, then

$$V_g = 1.2 \times 11.30 = 13.56\ \text{ft/sec}$$

$$q_{t,water} = 1.2 \times 0.575 = 0.690\ \text{MMscf/D}$$

   For this example, the well is above critical considering Coleman et al. ($.6 > .575$ MMscf/D) but below critical ($.6 < .69$ MMscf/D) according to Turner et al. We would say it is above critical because the more recent lower wellhead pressure correlation of Coleman et al. says it is flowing above critical. This example illustrates that the more recent Coleman et al. relationships require less flow to be above critical when analyzing data with lower wellhead pressures. Also, the example shows that the relationships require surface tension, gas density at a particular temperature, and pressure including use of a correct compressibility factor and gas gravity. If these factors are not taken into account for each individual calculation, then the approximate equation may be used. For this example, the already simplified Coleman et al. equation gives:

$$q_{t,water} = \frac{.0742 P d_{ti}^2}{(T+460)Z} \frac{(67-.0031P)^{1/4}}{(.0031P)^{1/2}}$$

$$= \frac{.0742 \times 400 \times 1.995^2}{(120+460) \times 0.9} \frac{(67-.0031 \times 400)^{1/4}}{(.0031 \times 400)^{1/2}}$$

$$= .579 \text{ MMscf/D}$$

and is very close to the previously calculated 0.575 MMscf/D.

3.2.5 Critical Flow Nomographs

To simplify the process for field use, the following simplified chart from Trammel and Praisnar[3] can be used for both water and condensate production. To use the chart, enter with the flowing surface tubing pressure (see the dotted line) at the bottom X axis for water and top axis for condensate. Move upward to the correct tubing size then either left for water or right for condensate to the required minimum critical flow rate.

Example 3.2: Critical Velocity from Figure 3-2

200 psi wellhead pressure
2 3/8-inch tubing, 1.995 inch ID
What is minimum production according to the Turner equations?

The example indicated by the dotted line shows that for a well having a wellhead pressure of 200 psi and 2 3/8-inch tubing, the flow rate must be at least ≈586 MMscf/D (actually 577 calculated) or else liquid loading will likely occur.

A similar chart was developed by Coleman et al. using the Turner et al. correlation for flowing wellhead pressures below approximately 800 psi (Figure 3-3). One set of curves is represented on this chart to be used for both water and condensate. The chart is used in the same manner as the above chart with no distinction between water and condensate. If water and condensate are present, the more conservative water coefficients are used anyway.

The Coleman et al. correlation would then be applicable for flowing surface tubing pressures below approximately 800 psi and the Turner chart (or Turner correlation) for surface tubing pressures above approximately 800 psi. The dividing line between using Turner

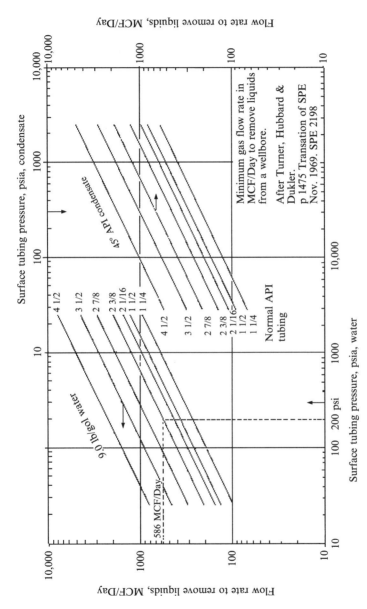

Figure 3-2. Nomograph for critical rate for water or condensate [after Reference 3] for a constant $Z = 0.9$, temperature of 60 °F, and the original Turner et al. assumptions of surface tension of $\sigma = 20$ dyne/cm for condensate, 60 dyne/cm for water, $\rho = 45$ lbm/ft³ for condensate, and 67 lbm/ft³ for water, and gas gravity $= 0.6$.

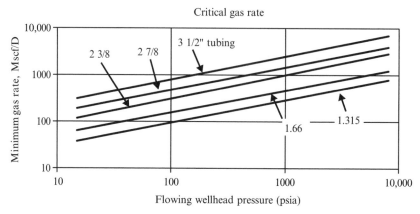

Figure 3-3. The Coleman et al. nomograph for critical rate[2] (for lower surface tubing pressures).

or Coleman might best be obtained from experience or even a blend of the two in the interval of 500–1000 psi.

The chart in Figure 3-4 is another way to look at critical velocity. Although it was prepared using a routine calculating actual Z factor (gas compressibility) at each point, it still depends on fluid properties and temperatures. For this, 60 dyne/cm for surface tension, 67 lbm/ft^3, gas gravity of 0.6, and 120 °F were used

---

Example 3.3: Critical Velocity with Water: Use Turner et al.'s Equations with Figure 3-4

100 psi wellhead pressure
2 3/8-inch tubing, 1.995-inch ID
Read from Figure 3-4 a required rate of approximately 355 Mscf/D
Compare to the simplified Turner et al. equations using $Z = 0.9$ for simplicity.

$$V_{t,water} = \frac{5.32(67 - .0031P)^{1/4}}{(.0031P)^{1/2}}$$

$$= \frac{5.32(67 - .0031 \times 100)^{1/4}}{(.0031 \times 100)^{1/2}} = \frac{5.32 \times 2.86}{.557} = 27.22 \text{ ft/sec}$$

$$q_g = \frac{3.067 P V_g A}{(T+460)Z} = \frac{3.067 \times 100 \times 27.22 \times .0217}{580 \times .9} = 0.346 \text{ MMscf/D}$$

In this case, a difference between the calculations and reading from chart can be attributed to that fact that the chart was calculated using actual Z factors and not an assumed value of 0.9.

Using one of the critical velocity relationships, the critical rate for a given tubing size vs. tubing diameter can be generated as in Figure 3-5 where a surface pressure of 200 psi and a surface temperature of 80 °F are used. (In this case, specific liquid and gas properties were used in the critical flow equations rather than the typical values given above.) This type of presentation provides a ready reference for maximum tubing size given a particular well flow rate.

A large tubing size may exhibit below critical flow, and a smaller tubing size may indicate that the velocity will increase to be above critical. Tubing sizes approaching and less than 1 inch, however, are not generally recommended because they can be difficult to initially unload due to the high hydrostatic pressures exerted on the formation with small amounts of liquid. It is difficult to remove a slug of liquid in a small conduit. See also Bizanti and Moonesan[4] for pressure, temperature, and diameter relationships for unloading, and Nosseir et al.[5] for

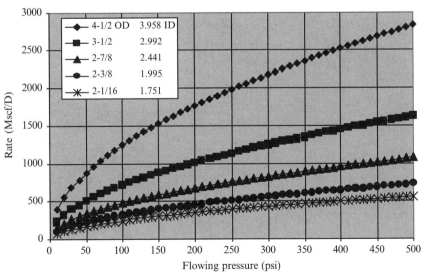

FIGURE 3-4. Simplified Turner critical rate chart.

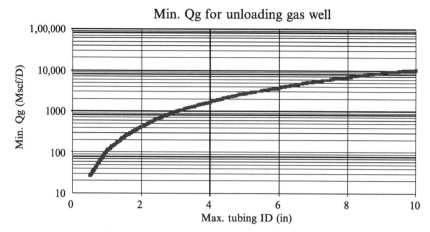

Figure 3-5. Critical rate vs. tubing size (200 psi and 80 °F) from Maurer Engineering, PROMOD program. Use this type of presentation with critical velocity model desired.

consideration of flow conditions leading to different flow regimes for critical velocity considerations.

3.3 CRITICAL VELOCITY AT DEPTH

Although these formulas are usually applied using the surface pressure and temperature, their theoretical basis allows them to be applied anywhere in the wellbore if pressure and temperature are known. The formulas are also intended to be applied to sections of the wellbore that have a constant tubing diameter. Gas wells can be designed with tapered tubing strings or with the tubing hung off in the well far above the perforations. In these cases, it is important to analyze gas well liquid-loading tendencies at locations in the wellbore where the production velocities are lowest.

For example, in wells equipped with tapered strings, the bottom of each taper size would exhibit the lowest production velocity and thereby would be first to load with liquids. Similarly, for wells having the tubing string hung well above the perforations, the analysis must be performed using the casing diameter near the bottom of the well because this would be the most likely location for initial liquid buildup. In practice, liquid-loading calculations should be performed at all sections of the tubing where diameter changes occur. For a constant diameter string, however,

Liquid transport in a vertical gas well

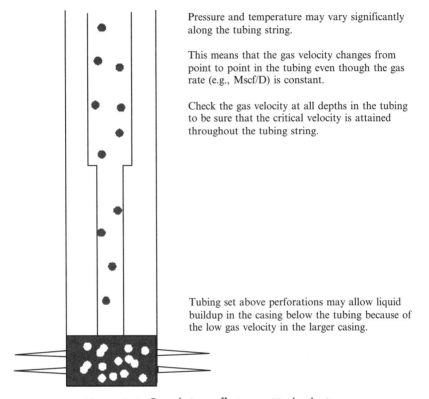

Pressure and temperature may vary significantly along the tubing string.

This means that the gas velocity changes from point to point in the tubing even though the gas rate (e.g., Mscf/D) is constant.

Check the gas velocity at all depths in the tubing to be sure that the critical velocity is attained throughout the tubing string.

Tubing set above perforations may allow liquid buildup in the casing below the tubing because of the low gas velocity in the larger casing.

Figure 3-6. Completions effect on critical velocity.

if the critical velocity is exceeded at the bottom of the string, then it will be exceeded everywhere in the tubing string.

In addition, when calculating critical velocities in downhole sections of the tubing or casing, downhole pressures and temperatures must be used. Minimum critical velocity calculations are less sensitive to temperature, which can be estimated using linear gradients. Downhole pressures, on the other hand, must be calculated by using flowing gradient routines (Nodal Analysis)™ or by using a gradient curve. The accuracy of the critical velocity prediction depends on the accuracy of the predicted flowing gradient. This could lead to problems if the gradient correlation predicted conditions indicative of one flow regime and the critical velocity calculated compared to the actual velocity indicated another flow regime.

3.4 CRITICAL VELOCITY IN HORIZONTAL WELL FLOW

The previous correlations for critical velocity cannot be used in inclined or horizontal wells. In deviated wellbores, the liquid droplets have very short distances to fall or rise before contacting the flow conduit, thus possibly rendering the mist flow analysis ineffective. Because of this phenomenon, calculating gas rates to keep liquid droplets suspended and maintaining mist flow in horizontal sections is a different situation than for tubing. Fortunately, hydrostatic pressure losses are minimal along the lateral section of the well and only begin to come into play as the well turns vertical where critical flow concepts again become applicable.

Another less understood effect that liquids can have on the performance of a horizontal well involves the geometry of the lateral section of the wellbore. Horizontal laterals are rarely straight. Typically, the wellbores "undulate" up and down throughout the entire lateral section. These undulations tend to trap liquid, causing restrictions that add pressure drop within the lateral. Several two phase flow correlations, which calculate the flow characteristics within undulating pipe, have been developed over the years and, in general, have been met with good acceptance. One such correlation is the Beggs and Brill method.[6] These correlations have the ability to account for elevation changes, pipe roughness and dimensions, liquid holdup, and fluid properties.

A rule of thumb developed from gas distribution studies suggests that when the superficial gas velocity (superficial gas velocity = total in situ

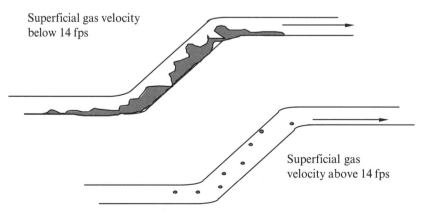

Superficial gas velocity below 14 fps

Superficial gas velocity above 14 fps

Figure 3-7. Effect of critical velocity in horizontal/inclined flow.

gas rate/total flow area) is in excess of ≈14 fps, then liquids are swept from low lying pipe sections as illustrated in Figure 3-7.

On examination, this is a conservative condition and requires a fairly high flow rate. Keep in mind, however, when performing these calculations that the gas velocity at the toe of the horizontal section can be substantially less than at the heel.

REFERENCES

1. Turner, R. G., Hubbard, M. G., and Dukler, A. E., "Analysis and Prediction of Minimum Flow Rate for the Continuous Removal of Liquids from Gas Wells," *Journal of Petroleum Technology*, Nov. 1969, pp. 1475–1482.

2. Coleman, S. B., Clay, H. B., McCurdy, D. G., and Norris, H. L., III, "A New Look at Predicting Gas-Well Load Up," *Journal of Petroleum Technology*, March 1991, pp. 329–333.

3. Trammel, P., and Praisnar, A., "Continuous Removal of Liquids from Gas Wells by Use of Gas Lift," SWPSC, Lubbock, Texas, 1976, p. 139.

4. Bizanti, M. S., and Moonesan, A., "How to Determine Minimum Flowrate for Liquid Removal," *World Oil*, September 1989, pp. 71–73.

5. Nosseir, M. A. et al., "A New Approach for Accurate Predication of Loading in Gas Wells Under Different Flowing Conditions," SPE 37408, presented at the 1997 Middle East Oil Show in Bahrain, March 15–18, 1997.

6. Beggs, H. D., and Brill, J. P., "A Study of Two-Phase Flow in Inclined Pipes," *Journal of Petroleum Technology*, May 1973, p. 607.

CHAPTER 4

SYSTEMS NODAL ANALYSIS*

4.1 INTRODUCTION

A typical gas well may have to flow against many flow restrictions in order for the produced gas to reach the surface separator. The gas must flow (1) through the reservoir rock matrix, (2) then through the perforations and possible gravel pack, (3) possibly through a bottomhole standing valve, (4) through the tubing, (5) possibly a subsurface safety valve, and (6) through the surface flowline and flowline choke to the separator. Each of these components will have a flow-dependent pressure loss. A change in any of the well restrictions will affect the well production rate. To determine overall well performance, all components of the well must be considered as a unit or total system.

One useful tool for analyzing well performance is Systems Nodal Analysis. Nodal Analysis divides the total well system into two subsystems at a specific location called the "nodal point." One subsystem considers the inflow from the reservoir, through possible pressure drop components, and to the nodal point. The other subsystem considers the outflow system from some pressure on the surface down to the nodal point. For each subsystem, the pressure at the nodal point is calculated and plotted as two separate, independent pressure–rate curves.

The curve from the reservoir to the nodal point is called the "inflow curve," and the curve from the separator down to the nodal point is called the "outflow curve." The intersection of the inflow and outflow curves is the predicted operating point where the flow rate and pressure

---

* TM Macco Schlumberger.

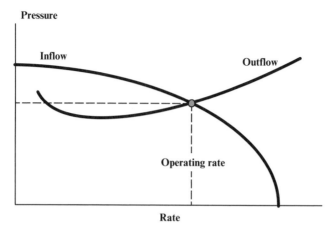

Figure 4-1. Systems Nodal Analysis.

from the two independent curves are equal. The inflow and outflow curves are illustrated in Figure 4-1.

Although the nodal point may be located at any point in the system, the most common position is at the mid-perforation depth inside the tubing. With this nodal point, the inflow curve represents the flow from the reservoir through the completions into the tubing, and the outflow curve represents the flow from the node to a surface pressure reference point (e.g., separator), summing pressure drops from the surface to the node at the midperforation depth.

The Nodal Analysis method uses single or multiphase flow correlations and correlations or theoretical models developed for the various components of reservoir, well completion, and surface equipment systems to calculate the pressure loss associated with each component in the system. This information is then used to evaluate well performance under a wide variety of conditions, which will lead to optimum single well completion and production practices. Nodal Analysis is useful for the analysis of the effects of liquid loading on gas wells.

Nodal Analysis will be illustrated as analyzing the effects of various tubing sizes on the ability of gas wells to produce reservoir liquids. Rough estimates of the onset of liquid-loading problems are possible, and examples will be given to illustrate the beneficial effects of reducing tubing size to increase the gas flow velocity in the tubing, thereby improving the efficiency of the liquid transport process.

Nodal Analysis is also used to clarify the detrimental effects of excessive surface production tubing pressure. Increased surface pressure adds backpressure on the reservoir at the sand face. The added backpressure reduces gas production and lowers the gas velocity in the tubing, which also reduces the efficiency with which the liquids are transported to the surface.

4.2 TUBING PERFORMANCE CURVE

The outflow or tubing performance curve (TPC) shows the relationship between the total tubing pressure drop and a surface pressure value, with the total flow rate. The tubing pressure drop is essentially the sum of the surface pressure, the hydrostatic pressure of the fluid column (composed of the liquid "holdup" or liquid accumulated in the tubing and the weight of the gas), and the frictional pressure loss resulting from the flow of the fluid out of the well. For very high flow rates, an additional "gas acceleration term" may be added to the pressure drop; however, it often is negligible compared with the friction and hydrostatic components. The frictional and hydrostatic components are shown by the dotted lines in Figure 4-2 for a gas well producing liquids. Duns and Ross[1] and Gray[2] are examples of correlations used for gas well pressure drops that include liquid effects.

Notice that the TPC passes through a minimum. To the right of the minimum, the total tubing pressure loss increases because of increased friction losses at the higher flow rates. The flow to the right of the minimum often is in the mist flow regime, which effectively transports small droplets of liquids to the surface.

At the far left of the TPC, the flow rate is low, and the total pressure loss is dominated by the hydrostatic pressure of the fluid column brought about by the liquid holdup or that percentage of the fluid column occupied by liquid. The flow regime exhibited in the far left-most portion of the curve is typically bubble flow, characteristically a flow regime that accounts for liquids accumulated in the wellbore.

Slightly to the left of the minimum in the TPC, the flow is often in the slug flow regime. In this regime, liquid is transported to the surface periodically in the form of large slugs. Fluid transport remains inefficient in this unstable regime as portions of the slugs "fall-back" as they rise and must be lifted again by the next slug. Fall-back and relifting the liquids results in a higher producing bottom hole pressure.

It is common to use the TPC alone, in the absence of up-to-date reservoir performance data, to predict gas well liquid-loading problems. It is generally believed that flow rates to the left of the minimum in the curve are unstable and prone to liquid-loading problems. Conversely, flow

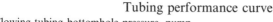

Tubing performance curve

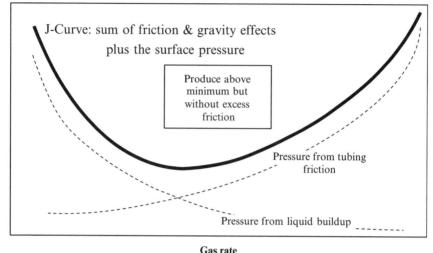

Figure 4-2. Tubing performance curve.

rates to the right of the minimum of the tubing performance curve
are considered to be stable and significantly high enough to effectively
transport produced liquids to the surface facilities.

Understandably, this method is inexact but is useful to predict liquid-
loading problems in the absence of better reservoir performance data.
Therefore, you can just select the flow rate you are measuring currently
and see if it is in a favorably predicted portion of the TPC or not, regardless
of having the reservoir inflow curve. Unfortunately, gas flow with liquid
correlations differ widely from one correlation to the next. This situation
is termed by some as the "biggest error in multiphase flow models."

With reservoir performance data, however, intersections of the tubing
outflow curve and the reservoir inflow curve allow a prediction of where
the well is flowing now and into the future if reservoir future IPR curves
can be generated.

4.3 RESERVOIR INFLOW PERFORMANCE RELATIONSHIP (IPR)

For a well to flow, there must be a pressure differential from the reservoir to the wellbore at the reservoir depth. If the wellbore pressure is
equal to the reservoir pressure, there can be no inflow. If the wellbore

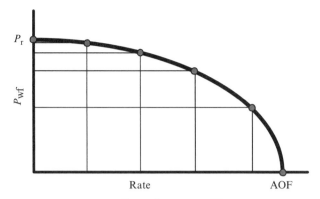

Figure 4-3. Typical reservoir IPR curve.

pressure at the pay zone is zero, the inflow would be the maximum possible—the Absolute Open Flow (AOF). For intermediate wellbore pressures, the inflow will vary. For each reservoir, there will be a unique relationship between the inflow rate and wellbore pressure.

Figure 4-3 shows the form of a typical gas well IPR curve. It is the deliverability curve, or "inflow performance relationship."

4.3.1 Gas Well Backpressure Equation

The equation for radial flow of gas in a well perfectly centered within the well drainage area with no rate dependent skin is:

$$q_{sc} = \frac{.000703\, k_g h (P_r^2 - P_{wf}^2)}{\mu Z T \ln(.472 \frac{r_e}{r_w} + S)} \tag{4-1}$$

where q_{sc} = gas flow rate, Mscf/D
k_g = effective permeability to gas, md
h = stratigraphic reservoir thickness (perpendicular to the reservoir layer), ft
P_r = average reservoir pressure, psia
P_{wf} = flowing wellbore pressure at the mid-perforation depth, psia
μ_g = gas viscosity, cp
Z = gas compressibility factor at reservoir temperature and pressure
T = reservoir temperature, °R

r_e = reservoir drainage radius, ft.
r_w = wellbore radius, ft.
S = total skin

Equation 4-1 can be used to generate an inflow curve of gas rate vs. P_{wf} for a gas well if all the above data are known. However, often the data required to use this equation are not well known, and a simplified equation is used to generate an inflow equation for gas flow that utilizes well test data to solve for the indicated constants.

$$q_{sc} = C(P_r^2 - P_{wf}^2)^n \tag{4-2}$$

where q_{sc} = gas flow rate, in consistent units with the constant C
n = a value which varies between approximately 0.5 and 1.0. For a value of 0.5, turbulent losses are indicated; for a value of 1.0, no turbulence losses are indicated.

This equation is often called the "backpressure" equation with the radial flow details of Equation 4-1 absorbed into the constant C. The exponent n must be determined empirically. The values of C and n are determined from well tests. At least two test rates are required because there are two unknowns, C and n, in the equation; four test rates are recommended to minimize the effects of measurement error.

If more than two test points are available, the data can be plotted on log-log paper and a least squares line fit is used to the data to determine n and C.

Taking the log of Equation 4-2 gives

$$\log(q_{sc}) = \log(C) + n\log(P_r^2 - P_{wf}^2) \tag{4-3}$$

On a log-log plot of rate vs. $(P_r^2 - P_{wf}^2)$, n is the slope of the plotted line and $\ln(C)$ is the Y-intercept, the value of q when $(P_r^2 - P_{wf}^2)$ is equal to 1. For two test points, the n value can be determined from the equation

$$n = \frac{\log(q_2) - \log(q_1)}{\log(P_r^2 - P_{wf}^2)_2 - \log(P_r^2 - P_{wf}^2)_1} \tag{4-4}$$

This equation may also be used for more than two test points by plotting the log-log data as described and picking two points from the best-fit line drawn through the plotted points. Values of the gas rate, q, and the corresponding values of $P_r^2 - P_{wf}^2$ can be read from the plotted line at the two points corresponding to the points 1 and 2 to allow solving for n.

Once n has been determined, the value of the performance coefficient C may be determined by the substitution of a corresponding set of values for q and $P_r^2 - P_{wf}^2$ into the backpressure equation.

If pseudostabilized data can be determined in a convenient time, then this equation can be developed from test data easily. Pseudo–steady state indicates that any changes have reached the boundary of the reservoir; but practically, it means for wells with moderate to high permeability that pressures and rates recorded appear to become constant with time. If the well has very low permeability, then pseudostabilized data may be nearly impossible to attain and then other means are required to estimate the inflow of the gas well. Rawlins and Schellhardt[3] provide more information on using the backpressure equation. For more details on the backpressure equation, see Appendix C.

4.3.2 Future IPR Curve with Backpressure Equation

For predicting backpressure curves at different shut-in pressures (at different times), the following approximation can be used for "future" inflow curves.

$$q_F = C_F (P_r^2 - P_{wf}^2)^n \tag{4-5}$$

where q_F = future gas rate

C_F = coefficient consistent with the gas rate and pressure units $= C_p \dfrac{(\mu_g z)_P}{(\mu_g z)_F}$

P_r = Average reservoir pressure, at current time, psia

P_{wf} = Flowing current wellbore pressure, psia

$(\mu_g z)$ = viscosity-compressibility product

subscript n is assumed constant; subscripts F and P signify "future" and "present" respectively

4.4 INTERSECTIONS OF THE TUBING CURVE AND THE DELIVERABILITY CURVE

Figures 4-4, 4-5, and 4-6 show a tubing performance curve intersecting a well deliverability inflow curve [inflow performance curve (IPR)]. These figures show the tubing curve intersecting the inflow curve at two places.

Tubing J-curve and flow stability

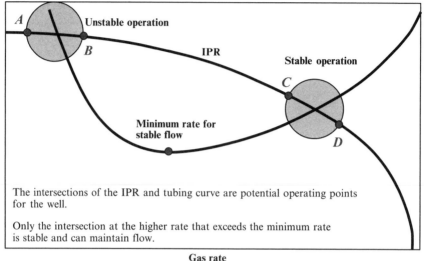

Figure 4-4. Tubing performance curve in relation to well deliverability curve.

Stability analysis shows that the intersection between points C and D is stable, whereas the intersection between A and B is unstable and, in fact, will not occur.

For example, if the flow rate strays to point D, then the pressure from the reservoir is at D but the pressure required to maintain the tubing flow is above D. The added backpressure against the sand face of the reservoir then decreases the flow back to the point of stability where the two curves cross. Similarly, if the flow temporarily decreases to point C, the pressure drop in the tubing is decreased, thus decreasing the pressure at the sand face and prompting an increase in flow rate back to the equilibrium point. Note also that the stable intersection between C and D is to the right of the minimum in the tubing curve. When the intersection of the tubing performance curve and the IPR curve occurs to the right of the minimum in the "J" curve, the flow tends to be more "stable," and stable vs. erratic flow usually means more production.

If, on the other hand, the flow happens to decrease to point A, the pressure on the reservoir is increased because of an excess of fluids accumulating in the tubing. The increase in reservoir pressure decreases the flow further, thus increasing the pressure on the reservoir further until the well dies. Similarly, if the well flows at point B, more pressure is available

Tubing J-curve and flow stability

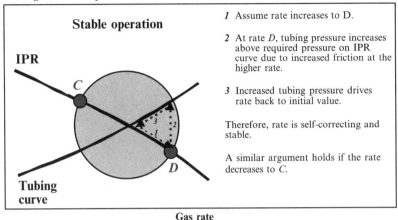

Figure 4-5. Stable flow.

Tubing J-curve and flow stability

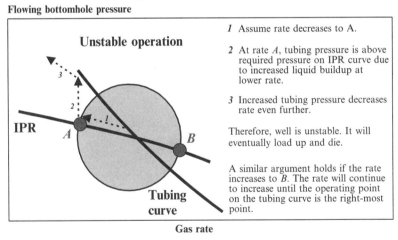

Figure 4-6. Unstable flow.

from the reservoir than needed by the tubing, so the rate moves to the right away from point B.

Thus, the crossing point of the IPR and the tubing performance curve to the right of the minimum of the "J" curve represents a stable flow condition. Flowing to the right of the minimum may be close to exceeding the "critical rate" but does not guarantee you will be above the "critical rate." To the left of the minimum represents unstable conditions where the well loads up with liquids and dies.

4.5 TUBING STABILITY AND FLOWPOINT

Another way to summarize unstable flow is presented in Figure 4-7. Here the difference between the two curves is the difference between the flowing bottomhole pressure and the flowing tubing surface pressure. The apex of the bottom curve is called the "flowpoint." Greene[4] provides additional information on the "flowpoint," as well as more information on gas well performance and fluid property effects on the AOF of the well.

The reasons for the flow rates below the "flowpoint" not being sustainable are explained at each rate by the slopes of the inflow and outflow curves as shown in Section 4.4. From Greene, "a change in the surface pressure is transmitted downhole as a similar pressure change, but a

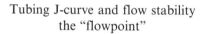

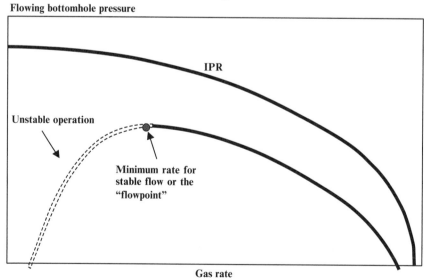

Figure 4-7. Flowpoint or minimum stable flow rate for gas well liquids production.

compatible inflow rate in the same direction as the pressure change does not exist." The result is an unstable flow condition that will either kill the well or, under certain conditions, move the flow rate to a compatible position above the "flowpoint" rate.

This discussion on stable rates is not the same as flowing below the "critical rate" as discussed in Chapter 3. There the mechanism is flowing below a certain velocity in the tubing, which permits droplets of liquid to fall and accumulate in the wellbore instead of rising with the flow. This discussion focuses on the interaction of the tubing performance with the inflow curve and also on reaching a point where the well will no longer flow in a stable condition.

The instability is brought on by regions of tubing flow where liquid is accumulating in the tubing because of insufficient gas velocity. Therefore, although the arguments are dissimilar, the root causes of each phenomenon are similar. Many Nodal programs will plot the "critical" point on the tubing performance curve. This point is often on the minimum of the tubing curve or to the right of the minimum in the tubing curve.

When analyzing a gas well, check for Nodal intersections that are "stable" and check for critical velocity at the top and bottom of the well. For instance, if selecting tubing size, choose one that will allow the well to flow above the "critical flow rate" and also to be "stable" from Nodal Analysis.

4.6 TIGHT GAS RESERVOIRS

A possible exception to the above stability analysis is the tight gas reservoir. A tight gas reservoir is generally defined as one where the reservoir permeability is less than 0.01 md. Tight gas reservoirs having low permeability have steep IPR relationships and react to changes in pressure very slowly. A possible tight gas inflow curve is shown in Figure 4-8. This figure shows that the right-most crossing of the tubing performance curve and the IPR might be to the left of the minimum of the "J" curve. The above "slope" arguments would lead to the conclusion that the right-most intersection is unstable but the well is flowing to the left of the minimum in the tubing performance curve. For tight gas wells, pseudo–steady state data take too long to obtain to get a good inflow curve, and often just using the "critical velocity" concept is the best tool to analyze liquid loading.

Tubing J-curve and flow stability
tight gas reservoir

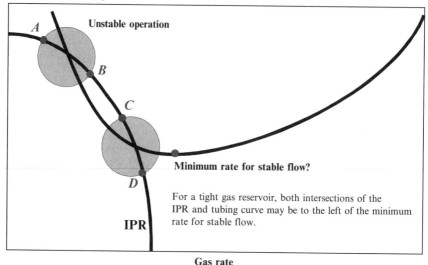

Figure 4-8. Tight gas well modeled with Nodal intersections.

4.7 NODAL EXAMPLE—TUBING SIZE

From the above analysis, it is clear that size (diameter) of the production tubing can play an important role in the effectiveness with which the well can produce liquids. Larger tubing sizes tend to have lower frictional pressure drops because of lower gas velocities, which, in turn, lower the liquid-carrying capacity. Smaller tubing sizes, on the other hand, have higher frictional losses but also higher gas velocities and provide better transport for the produced liquids. Chapter 5 provides additional information on sizing the tubing following this introductory example. In designing the tubing string, it is important to balance these effects for as long as possible. To optimize production, it may be necessary to reduce the tubing size later in the life of the well.

Figure 4-9 shows tubing performance curves superimposed over two IPR curves. For the higher pressure IPR curve, *C*, *D*, and *E* tubing curves would perform acceptably, but *D* and *E* would have more friction and less rate than would the tubing performance curve *C*. Curves *A* and *B*

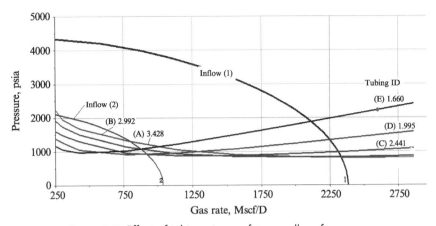

Figure 4-9. Effect of tubing size on future well performance.

may be intersecting to the left of the minimum in the tubing performance curve, and this is thought to generate unstable flow.

For the low pressure or "future" IPR curve, curves *A, B, C,* and *D* all intersect below the minimum for the tubing performance curves and as such would not be good choices. Tubing performance curve *E*, the smallest tubing, performs acceptably for the low pressure IPR curve. Curve *E* could intersect a little lower on the low pressure IPR curve but does not in this case because of a fairly high (400 psi) surface tubing pressure.

4.8 NODAL EXAMPLE—SURFACE PRESSURE EFFECTS: USE COMPRESSION TO LOWER SURFACE PRESSURE

Frequently, the production sales line pressure dictates the surface pressure at the wellhead, which may be beyond the control of the field production engineer. Some installations, however, have compressor stations near the sales line to maintain low pressures at the wellhead while boosting pressure to meet the levels of the sales line. Other methods to lower surface pressure are available to the engineer or technician. This section demonstrates the effects of lowering the wellhead pressure to enhance production and to better lift the produced liquids to the surface. Chapter 6 provides more information on the use of compression following this introductory example.

Figure 4-10 shows various tubing performance curves plotted against an IPR curve. The TPC curves or the "J" curves are all computed using the same tubing size but with various tubing surface pressures.

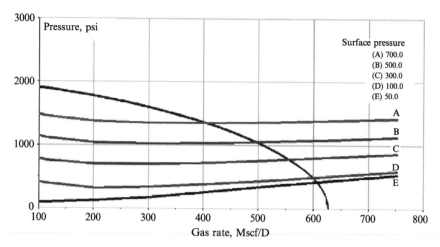

Figure 4-10. Effect of surface tubing pressure on well performances (23/8-inch to 19,000 ft).

Note that reducing the surface pressure lowers the tubing performance curve. Lower pressures are beneficial until the steep portion of the gas deliverability curve is reached, and then production returns diminish. For instance, the drop in surface pressure from 100 to 50 psi shows only a small gain in production because the deliverability curve is steep in this portion of the curve near the maximum flow rate or the AOF.

Reductions in the surface wellhead pressure can be implemented by:

- Compression
- Larger or "twinned" flowlines
- Elimination of small lines, bends, tees, elbows, chokes, or choke bodies at the surface
- Reduced separator pressure
- Eductors
- Removing scale, wax, or other solids from surface flowlines

4.9 SUMMARY NODAL EXAMPLE OF DEVELOPING IPR FROM TEST DATA WITH TUBING PERFORMANCE

This summary problem shows developing the gas IPR (inflow performance relationship) from test data and intersecting the IPR with a calculated tubing performance curve for a well. The object is to analyze if the well is, or if it will soon be, in any danger of liquid-loading problems.

Example Problem 4.9.1: Calculate the inflow curve from test data and intersect with tubing performance data.

Given data:

Reservoir Pressure, $\overline{P}_r$	3500 psia
2 7/8-inch tubing	2.441-inch ID
Depth (vertical well)	12000 ft
Water production	60 bbls-water/MMscf
Water Sp. Gr.	1.03
(Note: many operators do not know what the disposed water volume actually is)	
Gas gravity, γg	0.65
Tsurf	$120\,^{\circ}F$
BHT	$170\,^{\circ}F$
Psurf	300 psia

From flow-after-flow testing (Appendix C), the following pseudo–steady state data are available:

Gas Rate, q_g, Mscf/D	P_{wf}, psia	$(\overline{P}_r^2 - P_{wf}^2)^2 10^6$, psia
263	3170	2.911
380	2897	4.567
497	2440	7.006
640	2150	8.338

See Figure 4-11.
Solving for the "n" and "C" value for the "backpressure" equation:

$$n = \frac{\Delta\log(q_g)}{\Delta\log(\overline{P}_r - p_{wf}^2)} = \frac{\log730 - \log108}{\log10^7 - \log10^6} = \frac{2.863 - 2.033}{1} = 0.83$$

$$C = \frac{q}{(\overline{P}_r - P_{wf}^2)^n} = \frac{730}{(10 \times 10^6)^{0.83}} = 1.13 \times 10^{-3} \frac{\text{Mscf/D}}{\text{psia}^{2n}}$$

The inflow equation is then:

$$q_g = 1.13 \times 10^{-3} (3600^2 - p_{wf}^2)^{0.83}$$

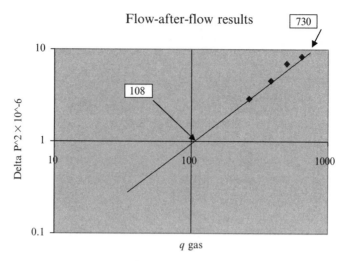

Figure 4-11. Log-log plot of flow-after-flow tests.

Using data from the gas backpressure equation, a conventional IPR equation can be plotted. Using a computer program, a tubing performance curve can be calculated using the Gray correlation (Appendix C). Plotting both the inflow and the tubing performance curve (outflow curve) gives the following plot. The pressure is the sum of the tubing surface pressure and the tubing pressure drop.

Tubing performance data using the Gray correlation:

q_g, Mscf/D	P_{wf}, psia
0	0
338	1200
524	900
666	865
772	835
846	838
890	841
905	844
1000	849
1200	855
1400	890

The following plot shows the intersection of the tubing performance curve at about 846 Mscf/D, and the intersection is to the right of the

System plot

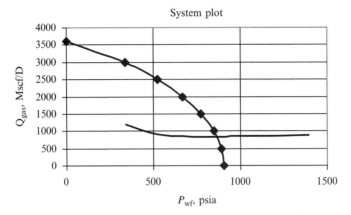

Figure 4-12. System plot of developed IPR curve intersected liquid/gas producing tubing performance curve.

minimum in the tubing curve so this should be a stable situation. However, because the minimum in the tubing curve is close to the inflow curve, further declines in the reservoir may lead to an unstable flow rate. The surface pressure is high for this well so a reduction in the surface tubing pressure would tend to allow flow further into the future if pressure declines; however, because the tubing curve is in the near vertical area of the IPR, a reduction in the surface tubing pressure would not increase flow much. See Figure 4-12.

The critical rate according to Coleman is calculated as follows with Z assumed as 0.9:

$$q_{t,water}(\text{MMscf}/\text{D}) = \frac{.0890 P d_{ti}^2}{(T + 460)Z} \frac{(67 - .0031P)^{1/4}}{(.0031P)^{1/2}}$$

$$= \frac{.0742 \times 300 \times 2.44^2 (67 - .0031 \times 300)^{1/4}}{580 \times .9(.0031 \times 300)^{1/2}}$$

$$= 0.751 \text{ MMscf}/\text{D}$$

The system plot shows a stable situation, and according to the Coleman et al. critical rate, the well is flowing above the critical rate of about 751 Mscf/D also.

4.10 SUMMARY

Systems Nodal Analysis can be used to study the effects of a wide variety of conditions on the performance of gas wells. The effects of tapered tubing strings, perforation density and size, formation fluid properties, and fluid production rates are just a few of the many parameters that the technique can analyze. Only a few sample problems varying tubing size and surface pressure with different inflow expressions are shown here.

Use Nodal Analysis to examine the effects of variables that you have control of, such as number of perforations, perhaps surface pressure, and tubular sizes, if designing a well or considering tubing re-size. For liquid loading, look for intersections of the tubing curve with the inflow curve to be to the right of the minimum in the tubing curve for stability. Use rates that will allow the well to flow above the critical rate at the surface and at points downhole as well.

REFERENCES

1. Duns, H., Jr., and Ross, N. C. J., "Vertical Flow of Gas and Liquid Mixtures in Wells," Proc. Sixth World Pet. Congress (1963), p. 451.

2. Gray, H. E., "Vertical Flow Correlations in Gas Wells," API User's Manual for API 14, "Subsurface Controlled Subsurface Safety Valve Sizing Computer Program," Appendix B, June 1974.

3. Rawlins, E. L., and Schellhardt, M. A., "Back Pressure Data on Natural Gas Wells and Their Application to Production Practices," Bureau of Mines Monograph 7, 1935.

4. Greene, W. R., "Analyzing the Performance of Gas Wells," presented at the annual SWPSC, Lubbock, Texas, April 21–22, 1978.

CHAPTER 5

SIZING TUBING

5.1 INTRODUCTION

As seen in Chapters 3 and 4, the size of the flow conduit through which the gas is produced (this could be the tubing or the casing-tubing annulus or simultaneous flow up the casing-tubing annulus and the tubing) determines the performance of velocity or siphon strings or just determines how well and for how long the production tubing will produce the well. The basic concept of tubing design is to have a large enough tubing diameter so that excessive friction will not occur and a small enough tubing diameter so that the velocity is high and liquid loading will not occur. The objective is to design a tubing installation that meets these requirements over the entire length of the tubing string or flow conduit. Also, it is desired to meet these requirements for as long as possible into the future before another well configuration may be required.

The concepts needed to properly size and evaluate a tubing change-out have been described in Chapter 4 using Nodal Analysis concepts and in Chapter 3 using critical velocity concepts. Both of these concepts should be considered when sizing tubing to reduce liquid loading. A well decline curve is also desirable to help decide if liquid loading is a problem and to properly evaluate the installation of smaller tubing.

5.2 ADVANTAGES AND DISADVANTAGES OF SMALLER TUBING

The reason to run smaller tubing is to increase the velocity for a given rate and to sweep the liquids out of the well and the tubing. In general, faster velocity reduces the liquid holdup (% liquid by volume in the

61

tubing) and lowers the flowing bottomhole pressure attributed to gravity effects of the fluids in the tubing. However, tubing too small for the production rate can cause excess friction and can create a larger flowing bottomhole pressure.

There are many other methods of de-liquefying a gas well, and tubing design must be compared to other possible methods before making a final decision. For instance, plunger lift is shown in Chapter 7 to work better, in general, in larger tubing. Therefore, you may reach a time in the life of the well when you must decide if you want to install and operate with smaller tubing or if you want to install plunger lift to reduce liquid loading in the future.

The advantages and disadvantages of smaller tubing should be evaluated before proceeding in this direction. Some of the disadvantages are:

1. Pressure bombs, test tools, and coiled tubing cannot be run in the smaller strings. This is especially true in 1.05-, 1.315-, 1.66-, and even in 1.9-inch OD tubing. This makes small diameter tubing unpopular with field personnel.
2. If you change to a smaller tubing today, then later you may have to downsize to even smaller tubing. There may be cases where using plunger lift could last longer into the future of the well without significant changes in the hardware. It is critical to evaluate the longevity of a smaller tubing design using Nodal Analysis or by comparison to the history of similar installations.
3. If the small tubing becomes loaded, then you cannot swab the tubing and may not even be able to nitrogen lift it. One-inch tubing is especially prone to load up and is hard to get started flowing again. Figure 5-1 shows how small tubing requires more pressure to support a given volume of fluid. The same volume of fluid that may be negligible in larger tubing can be significant in small tubing.

5.3 CONCEPTS REQUIRED TO SIZE SMALLER TUBING

To resize tubing, we need the reservoir inflow from a reservoir model or an IPR curve obtained from well test data. Then, we have the Nodal concepts of generating a tubing curve for various sizes of tubing, and we can obtain some information from the shape of the tubing curve. Also,

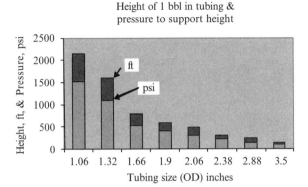

Figure 5-1. Effects of constant amount of liquid standing in various tubing sizes.

we have the concept of critical flow, and we want the velocity in the tubing to be greater than critical velocity so the holdup or percentage by volume of liquids in the tubing will be greatly reduced.

Example 5.1:

To illustrate these concepts, let's look at an example for various tubing sizes. Consider a well with these conditions:

Well Depth	10,000 ft
Bottomhole Temperature	180° F
Surface Flowing Temperature	80° F
Surface Flowing Pressure	100 psig
Gas Gravity	0.65
Water Gravity	1.02
Condensate Gravity	57 API
Water Rate	2 bbl/MMscf
Condensate Rate	10 bbl/MMscf
Reservoir Pressure	1000 psia
Reservoir Backpressure n	1.04
Reservoir Backpressure C	0.002 Mscf/D/psi^{2n}

The flowing bottomhole pressure for each tubing is calculated using the Gray correlation (see Appendix C) for a range of gas production rates and plotted on the same graph with the reservoir inflow curve (Inflow Performance Relationship [IPR]) in Figure 5-2.

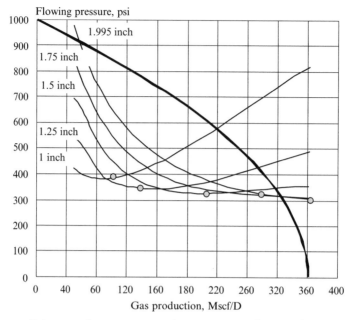

Figure 5-2. Tubing performances vs. tubing ID: Critical rates plotted on tubing curves.

From Figure 5-2 we see that:

- The 1-inch, the 1.25-inch, and the 1.5-inch ID tubing strings are acceptable because the minimum in the tubing or "outflow" curves is to the left of the expected intersection point with the IPR (or the point where they are calculated to flow).
- The 1.75-inch ID tubing curve is very flat at the intersection point, and we cannot be sure that the minimum is to the left of the intersection point of the IPR curve.
- The 1.995-inch ID curve definitely has the minimum somewhere to the right of the intersection point with the IPR curve.

We conclude from the Nodal plot that the 1.5-inch curve looks like the best design. The 1.75-inch performance is questionable and as the reservoir declines further, the 1.75-inch curve would definitely not be a good choice. The best design for the most production would be the 1.5-inch ID tubing for current conditions.

5.3.1 Critical Rate at Surface Conditions

Now let's check the critical rate for each tubing ID for tubing size. Because the surface pressure is low, we will use the Coleman et al.[2] findings for lower surface pressure wells, which modify the original Turner et al.[1] formulas. Because both water and condensate are present, we will conservatively use the water equation.

The surface critical gas rate required for water is calculated from Equation 5-1 using $Z = 0.9$ and is tabulated in Table 5-1:

$$Qgas, water = \frac{14.33PA(67 - .0031P)^{1/4}}{TZ(0.0031P)^{1/2}} \qquad (5\text{-}1)$$

The critical rates from surface pressures in Table 5-1 are plotted in Figure 5-2 as dots on the corresponding tubing curve. We can compare the critical rates for each tubing size with the flow rates predicted from the Nodal solution at the intersection point of each tubing curve with the IPR.

- The critical rate for the 1-inch ID tubing is to the right of the minimum in the tubing curve but is not close to the larger intersection of the tubing curve/inflow curve.
- The critical rate for the 1.25-inch tubing is perhaps a little to the left of the minimum in the tubing curve but is still to the left of the intersection.
- The critical rate for the 1.5-inch tubing seems just to the right of the minimum in the tubing curve but is still to the left of the intersection.
- The critical rate for the 1.75-inch tubing is to the left of the minimum in the curve but is still to the left of the intersection.
- The critical rate for the 1.995-inch curve is to the left of the minimum in the curve but is to the right of the tubing and IPR intersection. We have already stated not to use the 1.995-inch curve because the intersection with the deliverability curve is to the left of the minimum in the tubing curve, but the critical velocity also indicates to not to try to produce at this intersection because the critical rate of 346 Mscf/D is larger.

5.3.2 Critical Rate at Bottomhole Conditions

The previous analysis of critical rates made use of the well flowing surface pressure to calculate the critical rate at surface conditions. A similar analysis can be done at the bottomhole pressure conditions.

Table 5-1
**Bottomhole Critical Rates Needed at Nodal Intersections Compared
to Nodal Rates**

Tubing ID (in)	1.000	1.250	1.500	1.750	1.995
Nodal solution pressure (psia)	585	435	355	335	335
Nodal solution rate (Mscf/D)	220	275	320	325	325
Critical rate for Nodal solution pressure (Mscf/D)	167	226	294	388	505

Using the Nodal solution pressure (bottomhole pressure at the Nodal intersections), the Nodal solution rate can be calculated. If the critical rate calculated at the Nodal solution pressure is less than the Nodal solution rate, then the Nodal solution rates are acceptable; if not, then the critical velocity condition is violated.

Table 5-1 shows that the biggest tubing that has enough rate (above critical) at the bottom of the tubing is the 1.50" tubing. The larger ID tubings would have velocity at the bottom of the tubing that is less than the critical rate.

The calculation of critical rate at bottomhole conditions will depend somewhat on the particular method used to calculate the tubing curves. Multiphase flow correlations are developed for a range of fluid properties and tubing sizes that may not match particular well conditions.

Different multiphase flow correlations can often result in drastically different flowing gradients. The largest difference between correlations is usually in regard to how each calculates the beginning of the turn up or liquid loading at low rates. Thus, it is imperative to use a method appropriate for your well.

For lower rate gas wells with moderate liquids production, the Gray correlation (see Chapter 4) is good for predicting the tubing "J" curve and is recommended unless you have specific data that indicate otherwise. The Gray accumulation was used for the tubing curves in Figure 5-2.

The best way to ensure a good flowing bottomhole calculation is to measure the actual flowing bottomhole pressure and the associated well production rate and compare the different calculation methods to the measured data. Some software allows the user to adjust the calculations slightly to better match actual well data.

5.3.3 Summary of Tubing Design Concepts

When redesigning a tubing string:

• Check the Nodal Analysis for stability.

- Compare the Nodal solution rate to the critical velocity requirement at the top of the tubing.
- Compare the Nodal solution rate to the critical velocity at the bottom of the flow string. For a constant diameter string, if the velocity is above critical at the bottom of the string, then it will be acceptable for the entire length.
- Ensure that the flow correlation used to calculate the Nodal solutions is appropriate for your well conditions by comparison to some measured data if available.

In this example, the critical velocity at the bottom of the tubing limits the choices to the 1.5" ID tubing or smaller when considering critical velocity.

5.4 SIZING TUBING WITHOUT IPR INFORMATION

In the previous analysis, we used Nodal Analysis to evaluate different tubing options. This is the best method to design a tubing string as long as you have a good IPR curve. But, *you do not need to have an accurate representation of the reservoir or IPR curve* or *have to run a reservoir model* to make choices on the tubing size.

If you know where the well is flowing now, you can calculate the tubing curves for the current tubing string to see if you are currently flowing to the right or the left of the minimum in the tubing "J" curve. If you are flowing to the left of the minimum in the tubing curve, you can investigate different tubing sizes and generate curves where you would expect to flow to the right of the minimum curve. You can make these evaluations without having a reservoir curve (IPR curve) or without running a reservoir model.

If you do have a reservoir IPR curve to work with and the tubing curves intersect and match actual rates, then you can have more confidence in the results. But the reservoir curve is not necessary to analyze stability and critical velocity requirements.

Critical rates are typically evaluated at surface conditions. However, you can also calculate the downhole flowing pressures and enter the critical rate correlations for downhole conditions as shown in Table 5-1. You should especially make these calculations if you have any larger diameter flow paths, such as casing flow up to the entrance to the tubing. It is almost certain, however, that for wells on the verge of loading that flow up the casing will be below the critical velocity. However, if the length of casing flow from perforations to the tubing intake is not too long, then even

Table 5-2
Additional Field Results from Installation of Smaller Tubing (after Wesson[3])

			Typical Velocity String Results		
Well	Production String Size (in)	Perforation Depth (ft)	Initial Production	Velocity String Size (in)	Postproduction (in)
1	$2\frac{7}{8}$	8200	40 Mcfd 4 BLPD	$1\frac{1}{4}$	500 Mcfd 8 BLPD
2	$2\frac{7}{8}$	12,600	80 Mcfd 1–2 BLPD	$1\frac{1}{4}$	200 Mcfd 10 BLPD
3	$2\frac{7}{8}$	13,000	50 Mcfd 2 BLPD	$1\frac{1}{4}$	350 Mcfd 10 BLPD
4	$2\frac{7}{8}$	13,300	140 Mcfd 3 BLPD	$1\frac{1}{4}$	300 Mcfd 6 BLPD
5	$2\frac{7}{8}$	13,300	Dead	$1\frac{1}{4}$	250 Mcfd with soap injection
6	$2\frac{7}{8}$	11,380	150 Mcfd 6 BOPD	$1\frac{1}{4}$	155 Mcfd 12 BOPD
7	$3\frac{1}{2}$	11,860	8 Mcfd 2 BLPD	?	255 Mcfd 26 BLPD
8	$2\frac{7}{8}$	11,850	25 Mcfd 4 BOPD	$1\frac{1}{4}$	419 Mcfd 19 BOPD
9	$2\frac{7}{8}$	11,365	350 Mcfd 50 BWPD	$3865'-1\frac{1}{4}$ $7500'-1\frac{1}{2}$	450 Mcfd 115 BWPD
10	$2\frac{7}{8}$	9475	167 Mcfd 2 BOPD	$1\frac{1}{4}$	533 Mcfd 5 BOPD
11	$2\frac{7}{8}$	9415	167 Mcfd No liquid	$1\frac{1}{4}$	367 Mcfd 2 BLPD
12	$2\frac{7}{8}$	16,250	100 Mcfd No liquid	$1\frac{1}{2}$	425 Mcfd 3 BLPD
13	$3\frac{1}{2}$	14,900	440 Mcfd 35 BWPD	$4900'-1\frac{3}{4}$ $10000-2$	750 Mcfd 50 BWPD
14	$3\frac{1}{2}$	12,938	250 Mcfd 1 BWPD	$4538'-1\frac{1}{2}$ $8400'-1\frac{3}{4}$	575 Mcfd 2.5 BWPD

if it is flowing below critical, the net additional pressure drop may not be excessive. A Nodal program that can model flow string diameter changes with depth can analyze this situation. You should check downhole critical flow as a precaution even if the tubing size is constant down to the perforations.

5.5 FIELD EXAMPLE NO. 1—RESULTS OF TUBING CHANGEOUT

Regardless of the precautions listed in this section, there are many success stories related to coiled tubing and smaller tubing installations as described in Table 5-2. Still, economics must be considered, and one must be careful to consider whether a velocity string is the best method for long-term results. Other methods should be investigated to see if they would provide similar or greater rate benefits and perhaps require fewer modifications to the well over a time period.

5.6 FIELD EXAMPLE NO. 2—RESULTS OF TUBING CHANGEOUT

Dowell/Schlumberger published results from several case histories involving smaller tubing. A summary of some cases is shown in Table 5-3 from the Dowell/Schlumberger report.[4]

A typical production chart from the Dowell/Schlumberger report is reproduced in Figure 5-3. Clearly, dramatic improvements can be made by the proper and timely installation of coiled tubing in wells experiencing liquid loading.

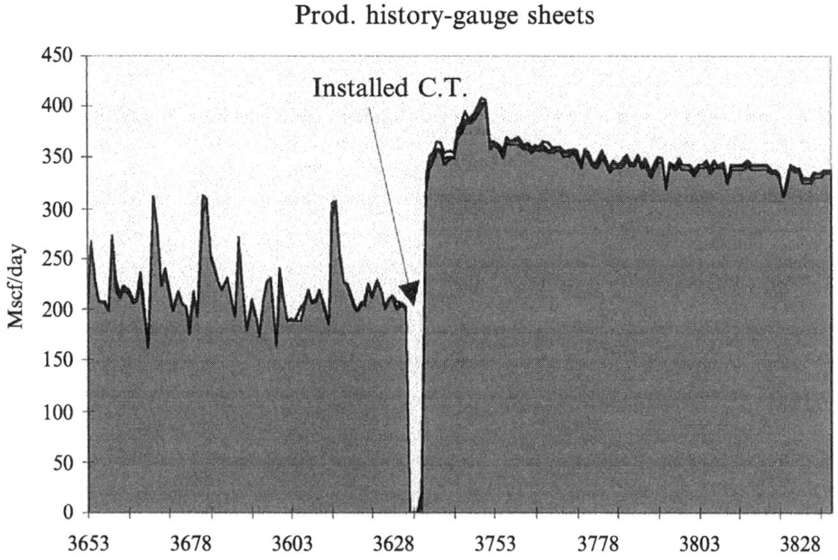

Figure 5-3. Example of rate change after coiled tubing installation.[4]

Table 5-3
Coiled Tubing Installation Results[4]

Well Name	Tbg OD"	CT OD"	Feet to Perfs	BHP (psi)	Mscf/D Prior CT	Mscf/D With CT	NPV ($)	Time to Payout Days	Other Information and Benefits
Well 1	$3\frac{1}{2}$	$1\frac{1}{4}$	12,700	1400	220	340	28200	34	Pulled compressor rates stabilize after job
Well 2	2 7/8	$1\frac{1}{2}$	14,200	750	400	390	11406	119	Good example—rates stabilize after job
Well 3	2 7/8	$1\frac{1}{2}$	15,360	1200	200	400	6900	90	Simplified well ops after job. Good production response
Well 4	2 7/8	$1\frac{1}{4}$	13,500	3400	185	175	23266	98	Delay compressor inst. rates stabilize after job
Well 5	2 7/8	$1\frac{1}{2}$	9,430	1100	625	550	14345	123	Delay compressor inst. rates stabilize after job
Well 6	2 7/8	$1\frac{1}{4}$	12,390	1700	370	360	11917	124	Delays compressor inst. simplifies well ops after job
Well 7	$3\frac{1}{2}$	$1\frac{1}{2}$	12,580	500	450	560	23141	96	Rates stabilize well
Well 8	2 7/8	$1\frac{1}{4}$	12,600	400	80	180	13534	132	Well almost dead before job. Rates stabilize
Well 9	2 7/8	$1\frac{1}{2}$	16,380	1400	280	170	−5499	DNP O	CT well above perfs (270' of 5")
Well 10	$3\frac{1}{2}$	$1\frac{1}{2}$	14,250	400	450	280	−32000	DNP O	Gauge sheets did not show need for CT. CT above perfs (160')
Summary			13339	1225	326	341	9521	102	

5.7 PRE- AND POST-EVALUATION

Another method used to evaluate a prospect for smaller tubing is the decline curve. Although this does not determine what size tubing should be used, it does show if the production is sharply dropping off. From the previous discussions of Nodal and Critical Velocity concepts, you can then analyze the well to see if decline dropoffs are a result of liquid loading. You might run a pressure bomb in the well to see if the well has liquid loading near the bottom of the well. Early remedial action will eliminate some problems, and any actions taken later will not show as quite a dramatic effect on production.

Also, check for holes in the tubing before making any well evaluation. This is especially true if the well has no packer because some liquids can fall back and reload the tubing. If there is a packer, then a hole will just allow the casing to pressure up, which could be a casing integrity problem or a corrosion problem.

Figure 5-4 shows the results of a study for a 10,000-foot well (after Weeks[5]).

Note that in late 1978 and early 1979, a sharp production decline was evident. It was determined that this sharp decline was caused by liquids

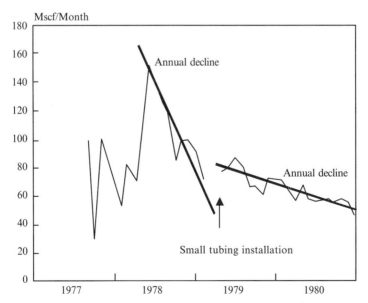

Figure 5-4. Example of slope change in decline change after coiled tubing installation.[5]

loading. A small string of coiled tubing was installed, and the shallower decline curve was measured after installation.

The production increased slightly after the installation of the small diameter tubing. If the decline curve information was not available, then the installer of the small diameter tubing might have thought that the installation did not cause a very favorable production response from the well. The point is that after installation of small tubing (or even plunger, gaslift, foam, or other methods), the production may or may not increase very much. However, if the steep decline curve is arrested and a flatter decline with less interference from liquids loading is achieved, then the installation is a success, and more recoverable reserves will result.

Therefore, always try to get decline curve information before installation of small tubing (or other methods of dewatering) and then keep post-installation decline curve data to properly evaluate the installation.

Figure 5-5 shows the completion corresponding to the decline curve in Figure 5-4.

Although 1-inch coiled tubing was (initially) successful in this case history, use 1-inch string with caution. When the tubing is this small, an intermittent slug of liquid can load the tubing, and it can be difficult or impossible to get the string unloaded again.

One other caution needs to be considered when viewing decline curves as indicated by the decline curve of Figure 5-6. At first glance, this curve appears to be a fairly normal decline curve without a sharp break in the curve to a steeper decline. One might conclude that this well has no liquid loading problems.

However, after some diagnosis, this well, even though it has a smooth downward decline, was found to be liquid loaded. In fact, it was liquid loaded from the first day of production. It has a smooth decline because it is always liquid loaded and did not show the characteristic change in decline rate from no loading to liquid loading conditions. Because a decline rate change was not observed, it was assumed that the well was not liquid loaded. This well is capable of producing a higher rate on a shallower decline curve than the one shown in Figure 5-6.

5.8 WHERE TO SET THE TUBING

It is recommended to set the tubing at the top of the pay but not below approximately the top one-third of pay. If the tubing is set too deep, liquid could collect over the perforations during a shut-in. When

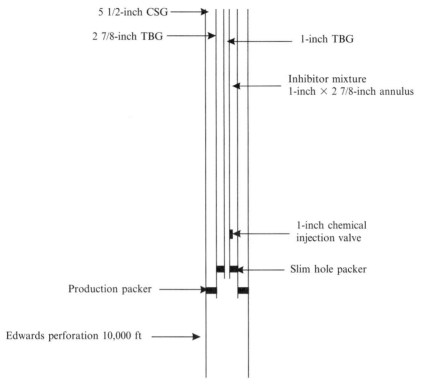

Figure 5-5. Completion[5] used to generate data for Figure 5-4.

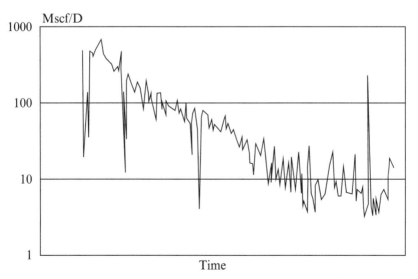

Figure 5-6. Rate vs. time: Well that is liquid loaded.[3]

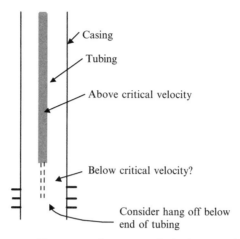

Figure 5-7. Illustration of setting end of tubing too high.

the well is brought back on production, the relatively large liquid volume in the casing-tubing annulus must be displaced into the tubing making the well difficult, if not impossible, to flow because of a high fluid level in the tubing. Also, if the tubing end is set below the perforations, then pressure accumulation during shut-in cannot push liquids below the tubing end or near the tubing end because there is no place for the liquids to enter the formation. You could set below the upper one-third of the pay zone if you are fairly sure perforations are open below where tubing end is landed. If you know perforations are open near the bottom of the pay, you could land the end of the tubing deeper.

5.9 HANGING OFF SMALLER TUBING FROM THE CURRENT TUBING

Sometimes tubing is landed high (Figure 5-7), and the flow through the casing below the tubing end is most likely well below critical, creating an extra pressure drop because of liquid loading of the casing.

If wells are completed with considerable casing flow between the perforations and the tubing intake or if there is a very large pay interval and the tubing is currently set at the top of the pay, it may be beneficial to hang off a section of smaller tubing from the end of the current tubing end to a deeper well depth. There are at least two systems[6,7] that will allow you to hang off a smaller tubing from the end of the current tubing.

The two tools are the double grip hydraulic set and wireline set packer for suspending coiled tubing.[7] The packer can be set hydraulically (Figure 5-8) or by a charge using an electric line—set similar to the Baker Model D packer. For the hydraulically set packer, there is a stinger torn disconnecting the CT above the packer after it is set.

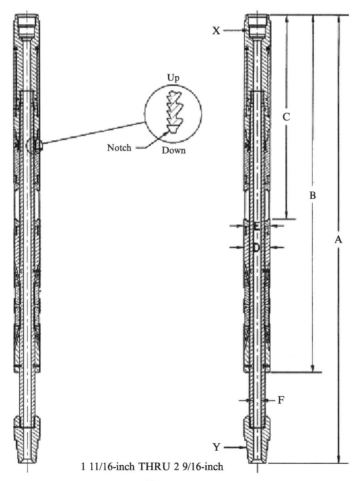

Figure 5-8. A hydraulic set packer[7] that can be run inside existing tubing to hang off a smaller section of coiled tubing to eliminate areas of flow below critical velocity below high set tubing.

5.10 SUMMARY

Smaller tubing can be successfully used at rates of several hundred Mscf/D as opposed to smaller rates where plunger lift might be used. It has fewer problems if the tubing installed is well above 1" ID considering limitations on tools that can be run and methods used to possibly unload the well.

- Size the tubing using Nodal Analysis for stability and critical rate for minimum velocity and use the conservative higher rates of the two indicated for a particular tubing size.
- Be sure that the tubing size selected flows above critical velocity from top to bottom and that the tubing is landed so that no large tubulars below the tubing bottom contribute to liquid loading.
- Do not land the tubing below the perforations but rather at the top or approximately in the top one-third of the pay to avoid large liquid slugs on startup.
- A packer will avoid annulus pressure cycling but planning ahead for possible plunger lift would dictate a completion without a packer.
- Even if the well is producing what seems to be an acceptable rate, check for liquid loading of flow below critical.
- Analyze any changes in tubing size not only from the immediate rate that is obtained but also from the slope of the new decline curve before conclusions are reached on the effects of a new completion.

REFERENCES

1. Turner, R. G., Hubbard, M. G., and Dukler, A. E., "Analysis and Prediction of Minimum Flow Rate for the Continuous Removal of Liquids from Gas Wells," *Journal of Petroleum Technology*, November 1969, pp. 1475–1482.

2. Coleman, S. B., Clay, H. B., McCurdy, D. G., and Norris, H. L., III, "A New Look at Predicting Gas-Well Load Up," *Journal of Petroleum Technology*, March 1991, pp. 329–333.

3. Wesson, H. R., "Coiled Tubing Velocity/Siphon String Design and Installation," 1st Annual Conference on Coiled Tubing Operations & Slimhole Drilling Practices, Adams Mark Hotel, Houston, TX, March 1–4, 1993.

4. WIS Solutions on "Coiled Tubing Velocity Strings: A Simple, Yet Effective Tool for the Future Technology," Schlumberger, Dowell.

5. Weeks, S. G., "Small Diameter Concentric Tubing Extends Economical Life of High Water-Sour Gas Edwards Producers," SPE 10254, presented at the 56th Annual Fall Technical Conference and Exhibition of the SPE of AIME, San Antonio, TX, October 5–7, 1961.

6. Campbell, J. A., and Bays, K., "Installation of 2 7/8-in. Coiled-Tubing Tailpipes in Live Gas Wells," OTC 7324 presented at the 25th Annual OTC in Houston, TX, May 3–6, 1993.

7. Petro-Tech Tools, Inc., Houston, TX.

CHAPTER 6

COMPRESSION

6.1 INTRODUCTION

In general, lowering the surface pressure of a well by using compression will result in more production. This is true for flowing wells and with nearly all lifted wells. Depending on the individual well, a well's production can be increased over a range from only a few percent to several times the current production of the well.

Some artificial lift methods, such as beam pumps, will respond to lower surface pressure with more production as tubing stretch is reduced, and lower casing pressure allows a higher fluid level in the annulus for the same rate. Then more production can be achieved.

For flowing gas wells, lowering the surface pressure can significantly increase the production, prolong the life of the well, and increase reserves.

Lowering the surface pressure of a gas well has two effects, both of which are beneficial for liquid loading.

- The flowing bottomhole pressure decreases, increasing the production rate and gas velocity throughout the wellbore.
- The required critical rate to remove liquids decreases because of the reduced pressures.

Wells flowing at just below the critical velocity before compression can be increased to rates beyond the critical velocity with compression, thereby alleviating liquid-loading problems. Often, compression provides an economic alternative to tubing changes for offshore wells where the workover costs can be prohibitive. Because the critical velocity is directly proportional to surface pressure (see Chapter 4), compression installations can be sized based on the critical velocity calculation. For example, the compression station can be sized based on a surface pressure that maintains the tubing production velocity at some percentage above the minimum critical velocity. With estimates of the decline of the reservoir, the installation size can be optimized to produce the most gas over the longest period.

6.2 NODAL EXAMPLE

Systems Nodal Analysis tools are ideally suited to analyze the effects of reducing the surface tubing pressure with compression. The technique allows for the simultaneous comparison of the effects of various surface pressures on well productivity. Figure 6-1 shows results from a sample calculation showing the effects that various surface pressures have on the production of a particular well. Several different surface pressures are plotted against a number of well inflow performance curves that define the decline of the reservoir. The current well performance is indicated by the top curve having the highest shut-in reservoir pressure.

The analysis shows that by installing a compressor to lower the surface pressure from 500 to 100 psi the well's production will increase from 1700 to approximately 2500 Mscf/D. This lower surface pressure will also prolong the life of the well and will increase reserves. At the 500-psi surface pressure, the well will likely begin to experience liquid loading problems

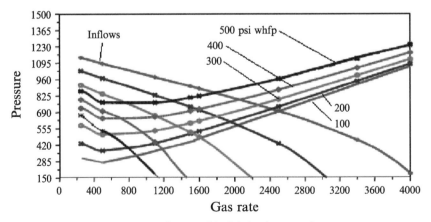

Figure 6-1. Systems Nodal Analysis results.

at a time corresponding to the third IPR curve where the intersection of the tubing and IPR curves occurs near the minimum of the tubing curve. At the 100 psi surface pressure, however, the well does not experience liquid-loading difficulties until sometime past the time of the last inflow curve.

6.3 COMPRESSION WITH A TIGHT GAS RESERVOIR

When compression is applied to tight gas reservoirs, it often results in a production increase, which often appears short-lived. This is illustrated Figure 6-2, which shows the effects of compression on a somewhat emphasized steep inflow curve for a tight gas reservoir.

The inflow curve, or deliverability curve, for tight gas reservoirs is characterized by a near vertical curve corresponding to the lowest drawdown pressures. As seen in Figure 6-2, lowering the surface pressure has little effect on the tight gas production rate because of the near vertical slope of the IPR at the point of intersection. The high perm reservoir would have a much more significant production increase.

Often, small increases in production are realized in a tight gas reservoir from compression; however, as the reservoir returns to pseudo–steady state conditions, the production returns to near its initial stable value as shown in the figure by the intersection of the tubing performance curve and the IPR.

As the reservoir declines, however, the net increase in production can be more significant as the "compressed" tubing performance curve will

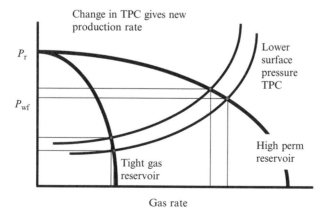

Figure 6-2. Tight gas reservoir example.

stay nearer maximum flow in the future, whereas tubing performance without compression will intersect the inflow in areas of liquid loading in the tubing.

The long-term cumulative production gain must be carefully analyzed to determine whether the expense of the compression station is warranted by the economics of increased reserves.

6.4 COMPRESSION WITH PLUNGER LIFT SYSTEMS

Plunger lift is one artificial lift system that can benefit greatly from compression. The basics of plunger lift are discussed in Chapter 7, where it is pointed out that lower pressure at the wellhead is very desirable. Both gas-powered and electric compressors have been shown to have application in plunger lift installations.

Figure 6-3 shows a simple schematic of a plunger lift installation equipped with a surface compressor. The compressor is switched to lower

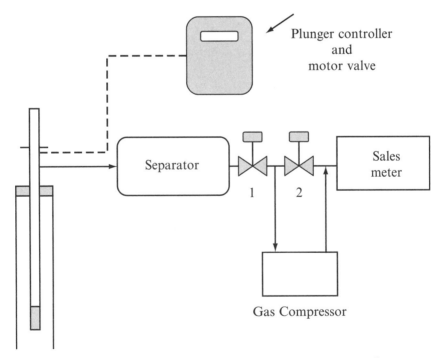

Figure 6-3. Compression system installation with plunger lift.[1]

the wellhead pressure when the well is open for production. The compressor can thus be valved to numerous, staged plunger lift wells, lowering the wellhead pressure during the production cycle then being valved off during the buildup cycle.

Plunger lift is sometimes thought of as an intermittent gas lift system where the energy is provided by the reservoir and not a compressor. The reservoir gas is sometimes supplemented by external gas. In general, compression to lower surface tubing pressure would, in almost every case, also help conventional intermittent or continuous gaslift.

Figure 6-4 shows production data for a plunger lift installation equipped with compression. The chart shows a dramatic increase in production after installing the surface compressor. The initial production followed a fairly steep decline until early in 1993 when it was put on compression; at which time, the oil and gas increased markedly.

Compression can also be used to inject gas into the casing of a plunger lift well to reduce the time required for the plunger to move to the surface. Morrow and Aversante[2] further discuss compression with plunger lift.

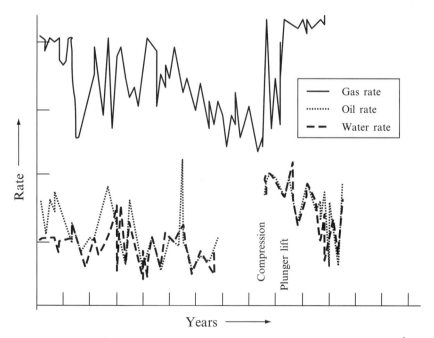

Figure 6-4. Performance improvement using plunger lift and compression.[1]

6.5 COMPRESSION WITH BEAM PUMPING SYSTEMS

In a beam pump system, the production is governed primarily by the downhole stroke, the number of strokes per minute (SPM), and the pump size. The wellhead pressure usually has only a minor influence on production. Lowering the wellhead pressure, however, does have some effect on the fluid level in the casing if the casing is open to the tubing at the surface as it is normally.

Figure 6-5 compares the effect of three wellhead pressures on the liquid level in the tubing/casing annulus. Note that these assume a constant production rate and flowing bottomhole pressure. The figure shows that as the wellhead pressure is reduced, the liquid level in the casing/tubing annulus is raised substantially. Conversely, high wellhead pressure puts the liquid level in the annulus low in the well near the pump intake. If the wellhead pressure is too high and the annulus fluid level too low, there exists a distinct possibility of incomplete pump fillage and lower pump efficiency.

Therefore, one way to ensure proper pump fillage and more efficient pump operation is to lower the surface wellhead pressure by compression. If the liquid level was initially low and pump fillage was incomplete, compression will increase the well's production. In addition, because the

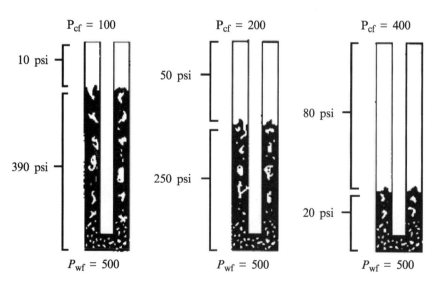

Figure 6-5. Pressure relations on a pumping well with a gaseous fluid column.[3]

higher liquid level in the annulus is available, the unit can also be run at higher speeds, which results in higher production rates. The higher speeds with good pump fillage will lower the fluid level and also lower the bottomhole pressure.

Increased pump fillage can also reduce pump failures by eliminating "fluid pound" as the plunger strikes a gas-liquid interface on the down-stroke. At a lower surface pressure, the fluid level comes up for the same rate. Then, the fluid level can be pumped down, increasing the rate. If the fluid level does not come up, then there is no fluid level to pump down.

6.6 COMPRESSION WITH ELECTRIC SUBMERSIBLE SYSTEMS

Electric submersible pumps (ESPs) operate with a fixed pressure between the pump intake and the pump discharge. This translates to a fixed pressure increase between the well's bottomhole flowing pressure and the wellhead pressure. Therefore, lowering the surface wellhead pressure on a typical ESP installation usually proportionally lowers the flowing bottomhole pressure. The lower bottomhole flowing pressure then takes backpressure away from the reservoir, which increases production and lowers the horsepower demand of the unit.

6.7 TYPES OF COMPRESSORS

Several compressor types are used to lower the pressure on entire fields of gas wells or to lower the pressure on individual gas wells.

For single-well applications, the following list of compressor types may be used.

- Rotary lobe
- Re-injected rotary lobe
- Rotary valve
- Liquid ring
- Liquid injected rotary screw
- Reciprocating
- Screw
- Sliding vane

The description for the first six types of equipment is taken from Thomas.[4]

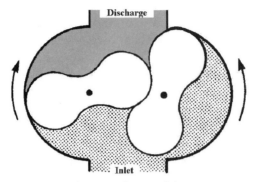

Figure 6-6. Elements of rotary lobe compressor.

6.7.1 Rotary Lobe Compressor (Figure 6-6)

- Low cost per cfm
- Air cooled
- Approximately 2.0 compression ratio
- Small amounts of liquid ingestion are acceptable
- High displacement is achievable (50–12,000 cfm)
- Power frame supporting bearings, gears, and shafts

6.7.2 Re-Injected Rotary Lobe Compressor

- Low cost per cfm
- Air cooled
- Approximately 4.0 compression ratio (high vacuum)
- Small amounts of liquid ingestion are acceptable
- High displacement is achievable (50–12,000 cfm)
- Power frame supporting bearings, gears, and shafts
- Requires intercooler

6.7.3 Rotary Vane Compressor

- Medium cost per cfm
- Liquid cooled (jacket)
- Approximately 4.5 compression ratio
- Medium displacement
- Power frame required
- Requires external oil lubrication system

- Once-through lubrication not recovered
- No amount of liquid ingestion

6.7.4 Liquid Ring Compressor

- Medium cost per cfm
- Liquid injected
- Approximately 4.0 compression ratio
- High displacement
- Power frame required
- Requires seal liquid cooling system (normally oil)
- Requires gas/liquid separator
- Large amounts of liquid ingestion possible (but water will contaminate oil system, requiring replacement of seal fluid)
- Generates about 25 psi delta pressure

6.7.5 Liquid Injected Rotary Screw Compressor (Figure 6-7)

- Higher cost per cfm
- Liquid injected
- Approximately 6.0 compression ratio
- Medium displacement
- Power frame required

Figure 6-7. Elements of a screw compressor.

- Requires seal oil cooling system
- Requires gas/oil separator
- Liquid ingestion dilutes seal oil
- Unless specifically designed for vacuum and low discharge pressure operation, then:
  - Questionable mechanical seals
  - High backpressure valve
  - High separator velocities can cause problems
- Can handle very high compression ratios in one stage of compression as the oil absorbs most of the heat of compression. Excellent for very low suction pressure, even down to vacuum. Oil cooling system required.
- Except for gear amplification, very few wearing parts, which provides very high reliability.
- Mechanical and adiabatic efficiency is high, if unit is run at design conditions.
- Efficiency suffers if unit is run too far off of initial design conditions or if multiple stages are used.
- Limited to 250 psig discharge pressure. Flexibility is limited to the initial design compression ratio of the rotors.
- Oil can become contaminated with heavy hydrocarbons and other liquids, causing operational problems. Selection of proper oil type is absolutely critical. Test oil frequently for fines content.

6.7.6 Reciprocating Compressor (Figure 6-8)

- High cost per cfm
- Air or liquid cooled
- Approximately 4.0 compression ratio
- Low displacement/power frame
- No amount of liquid ingestion allowed
- Valve losses greatly affect compression ratio and volumetric efficiency.
- Most flexible of all compressors because it can handle varying suction and discharge pressures and still maintain high mechanical and adiabatic efficiency.
- Overall compression ratio is dependent only on discharge temperature and rod load rating of frame. Units can be two staged (or even three staged) to produce very high discharge pressures with low suction pressure.

Figure 6-8. HP Gas Engine Drive Reciprocating Compressor Package. Operating conditions: Ps0/Pd50 Psig at 40 Mscf/D.

- Level of knowledge required for maintaining unit is low. Basically, any engine mechanic can be a good compressor mechanic.
- Most likely highest operating expense and downtime because of compressor valve maintenance. This valve maintenance is highly dependent on gas quality (solid and liquid contamination), which can be a problem with wellhead compression.
- Not efficient with low suction pressures

6.7.7 Sliding Vane Compressor

- Medium cost per cfm
- Liquid cooled (jacket)
- Approximately 3–4.5 compression ratio
- Medium displacement/power frame
- Requires external oil lubrication system
- Once-through lubrication; oil leaves with gas
- Tolerates no liquid ingestion
- Low capital and expense cost unit, very simple operation
- Simple design makes for easy and high availability (depending on quality)

- Useful in VRU service
- Bearings isolated from sour gas; separate lube system
- Blades wear on interior case, so compressor life is heavily dependent on gas quality and contaminants. Blades can get stuck in the case if many solids are present in the gas.
- Limited to lower discharge pressure and lower volume applications

6.8 GAS JET COMPRESSORS OR EDUCTORS

Gas jet compressors, or eductors, are classified as thermocompressors and are in the same family as jet pumps, sand blasters, and air ejectors. They use a high-pressure fluid (either gas or liquid) for motive power. Eductors using gas can impart up to two compression ratios; those using liquid can generate higher ratios.

The eductor, or gas jet compressor, operates on the Bernoulli principle as illustrated in Figure 6-9, which shows a jet pump (by Weatherford) operating with the same principles. The high-pressure motive fluid enters the nozzle and is accelerated to a high velocity/low pressure at the nozzle exit. The wellhead is exposed to the low pressure at the nozzle exit through the suction ports and is mixed with the motive fluid at the entrance to the throat. Momentum transfer between the motive and produced fluids in the throat and velocity decrease in the diffuser increases the pressure to the discharge pressure.

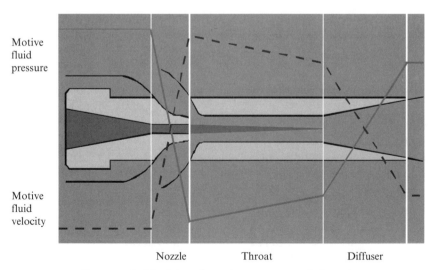

Figure 6-9. Principles of gas jet compressor (eductor).

Figure 6-10 shows a cross-section of a typical wellhead eductor. Figure 6-11 shows an eductor installed on a wellhead. Eductors have several advantages:

- No moving parts
- Low maintenance/high reliability

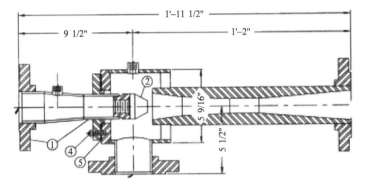

Figure 6-10. Cross-section of an eductor.[5]

Figure 6-11. Eductor installed on a wellhead.

Table 6-1
Single Well, Low Pressure, Low Volume Gas Well Compressor Operation
from a US Operator

Type	Cost (1000 US$)	Rate Mscf/D	Efficiency %	Discharge Psia	Suction Psia	CR
Reciprocating	40	200	83	50–60	0	3.5–4.0
Screw	30	150	60	5–30	16 in Hg	15
Vane	20	120–150	60	30	30	3
Liquid-Ring	12–15	150	60	22–30	5	5

Note: Re-injected lobe can be cheaper than Liquid-Ring.

- Easy to install, operate, and control
- Can handle liquid slugs
- Low initial cost/payback time usually short
- Nozzle sizes can be changed to meet changing well conditions

One successful configuration[5] uses a flooded screw compressor to pull the tubing/casing annulus down to 8–10 psig. A portion of the gas discharged by the compressor is used to drive an eductor to pull the tubing down to 1–5 psig. The exhaust of the eductor is combined with the casing gas and sent to the compressor.

An eductor sized to provide adequate velocity in 2-3/8" tubing to stay above the critical unloading rate requires less than 20 hp. This configuration has maintained nearly constant liquid levels for years without any additional lift.

Like some compressors, eductors do not have a minimum NPSHr (net positive suction head required) and are capable of moving 6–10 bbl/day.

The function of an eductor is to lower surface pressure using a power stream of fluid if available. If high-pressure fluid is available (e.g., from a nearby high-pressure gas well) to power the eductor, then why not take advantage of the stream to lower pressure on a lower rate well. Lower pressure means higher velocity, and higher velocity of gas will prevent liquids from accumulating in the flow string.

6.9 SUMMARY

Compression helps a liquid-loading well by increasing the gas velocity to equal or exceed the critical unload velocity and also lowers pressure on the formation for more production by lowering the wellhead flowing pressure.

There are several types of compressors with varying degrees of initial cost, operational cost and functionality for single-well compression.

Compression often is used for field wide compression to lower the gathering system pressure. Compression also helps other methods of artificial lift to different degrees. For instance, it may allow plunger lift to function where the plunger might not cycle without compression.

REFERENCES

1. Phillips, D., and Listiak, S., "Plunger Lifting Wells with Single Wellhead Compression," presented at the 43rd Southwestern Petroleum Short Course, Lubbock, Texas, April 23–25, 1996.

2. Morrow, S. J., and Aversante, O. L., "Plunger-Lift, Gas Assisted," 42nd Annual Southwestern Petroleum Short Course, Lubbock, Texas, April 19–20, 1995.

3. McCoy, J. N., Podio, A. L., and Huddleston, K. L., "Acoustic Determination of Producing Bottomhole Pressure," SPE Formation Evaluation, September 1988, pp. 617–621.

4. Thomas, F. A., "Low Pressure Compressor Applications," presented at the 49th Annual Liberal Gas Compressor Institute, April 4, 2001.

5. Simpson, D. A., "Use of an Eductor for Lifting Water," BP Forum on Gas Well De-Watering, Houstson, Texas, May 5, 2002.

CHAPTER 7

PLUNGER LIFT

7.1 INTRODUCTION

Plunger lift is an intermittent artificial lift method that usually uses only the energy of the reservoir to produce the liquids. A plunger is a free-traveling piston that fits within the production tubing and depends on well pressure to rise and solely on gravity to return to the bottom of the well. Figure 7-1 illustrates a typical plunger lift installation.

Plunger lift operates in a cyclic process with the well alternately flowing and shut-in. During the shut-in period with the plunger on the bottom, gas pressure accumulates in the annulus, and liquids have mostly already accumulated in the well during the last portion of the flow period. Liquids accumulate in the bottom of the tubing, and the plunger falls through the liquids to the bumper spring to await a pressure buildup period. The pressure of the annulus gas depends on the shut-in time, reservoir pressure, and permeability. When the annulus pressure increases sufficiently, the motor valve is opened to allow the well to flow. The annulus gas expands into the tubing, lifting the plunger and liquids to the surface, with some help from the producing gas.

The reservoir is allowed to produce gas until the production rate decreases to some value near the critical rate and until liquids begin to accumulate in the wellbore. The well is then closed, and the plunger falls back to the bumper spring—first through gas and then through some accumulated liquid.

The pressure buildup period follows. Then, using the gas pressure that has been allowed to accumulate in the annulus, the well is opened to production again, bringing the liquids and plunger to the surface. With the plunger at the surface, the well remains open, and the gas is allowed

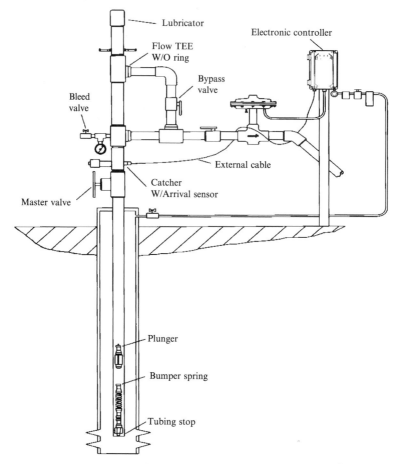

Figure 7-1. Typical conventional plunger lift installation.[1]

again to flow until production rates begin to fall. The well is closed in, and the plunger falls to the bottom, repeating the cycle.

Figure 7-2 shows an approximate depth-rate application chart where plunger lift is shown to be feasible in the region of lower rates and depths identified by the curve. This chart is an approximation, as plunger lift has been operated successfully to depths of 20,000 feet.

A plunger lift system is relatively simple and requires few components. A typical installation as in Figure 7-1 would include the following components:

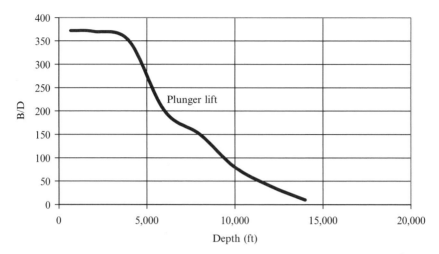

Figure 7-2. Approximate depth-rate application chart for conventional plunger lift.

- A downhole bumper spring, which is wirelined into the well to allow the plunger to land more softly downhole
- A plunger free to travel the length of the tubing
- A wellhead designed to catch the plunger and allow flow around the plunger
- A controlled motor valve that can open and close the production line
- A sensor on the tubing to sense arrival of the plunger
- An electronic controller that contains logic to decide how the cycles of flowing production and time of well shut-in period are determined for best production

7.2 PLUNGERS

Figure 7-3 shows some typical plungers that were tested to provide data for developing plunger lift system models.[1] The types shown are typical but do not include all types of plungers available to the industry. In this figure, the plungers are identified from left to right as:

1. Capillary plunger, which has a hole and orifice through it to allow gas to "lighten the liquid slug above the plunger"
2. Turbulent seal plunger with grooves to promote the "turbulent seal"
3. Brush plunger used especially when some solids or sand are present
4. Another type of brush plunger

5. Combination grooved plunger with a section of "wobble washers" to promote sealing
6. Plunger with a section of turbulent seal grooves and a section of spring-loaded expandable blades. Also, a rod can be seen that will open and close a flow-through path through the plunger depending if it is traveling down or up
7. Plunger with two sections of expandable blades with a rod to open flow through plunger on downstroke
8. Mini-plunger with expandable blades
9. Another with two sections of expandable blades and rod to open flow-through passage during plunger fall
10. Another with expandable blades and a rod to open a flow through passage during the plunger fall and to close it during the plunger rise
11. Wobble washer plunger and rod to open flow passage during the plunger fall
12. Plunger with expandable blades with a rod to open a flow-through passage on the plunger fall

Several of these plungers have a push rod to open a flow passage through the plunger to allow flow through the plunger when falling to increase the fall velocity. When the plunger arrives at the surface, the push rod forces the flow passage open for the next travel downward. When the plunger hits bottom, the rod is pushed upward to close the flow passage for the next upward travel.

In testing, the brush plunger was found to show the best seal for gas and liquids; however, it typically wears sooner than other plungers. The

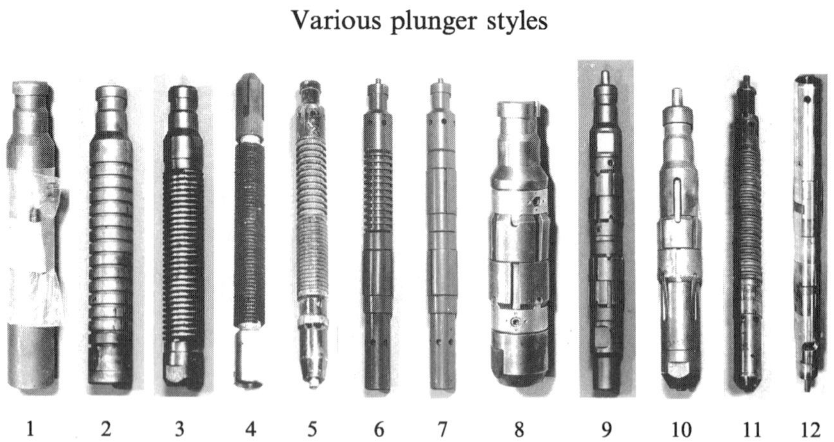

Various plunger styles

Figure 7-3. Various types of plungers.

brush plunger is the only plunger that will run in wells making a trace of sand or solids. Plungers with the spring-loaded expandable blades showed the second best sealing mechanism, and they do not wear nearly as fast as the brush plunger.

7.3 PLUNGER CYCLE

Plunger lift operates on a relatively simple cycle as illustrated in Figure 7-4. Figure 7-5 shows in more detail the casing and tubing and bottomhole pressures throughout one complete plunger cycle. The numbers on top of Figure 7-4 labeling the steps of the cycle are also provided on the figure for clarity.

1. The well is closed, and pressure in the casing is building. When the pressure is enough to lift the plunger and the liquids to the surface at a reasonable velocity (~750 fpm) against the surface pressure, the surface tubing valve will open.
2. The valve opens, and the plunger and liquid slug rise. The gas in the annulus expands into the tubing to provide the lifting pressure. Also, the well is producing some during the rise time to add to the energy required to lift the plunger and liquid.

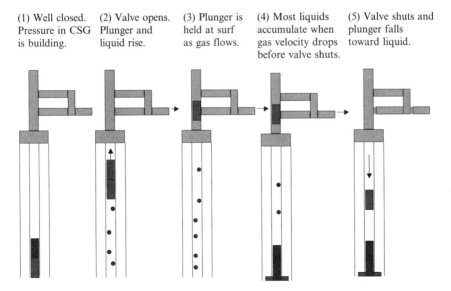

(1) Well closed. Pressure in CSG is building.   (2) Valve opens. Plunger and liquid rise.   (3) Plunger is held at surf as gas flows.   (4) Most liquids accumulate when gas velocity drops before valve shuts.   (5) Valve shuts and plunger falls toward liquid.

Figure 7-4. Simplified pictorial illustrations of plunger cycle events.

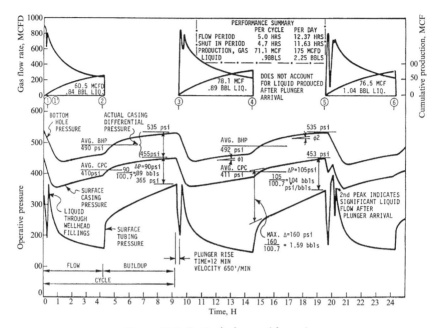

Figure 7-5. Typical plunger life cycle.

3. The liquid reaches the surface and travels down the flowline. The plunger is held at the surface by pressure and flow. The gas is allowed to flow.
4. The flow velocity begins to decrease, and liquids begin to accumulate in the bottom of the well. The casing pressure begins to rise, indicating a larger pressure drop in the tubing. If flow is allowed to continue too long, a "too large" liquid slug will accumulate in the bottom of the well, requiring a high casing buildup pressure to lift it.
5. The valve is shut, and the plunger falls. The liquids are mainly at the bottom of the well. The plunger will hit the bottom, and the cycle will repeat.

The cycles continue and may be adjusted according to different algorithms that may be programmed into the various controllers available.

7.4 PLUNGER LIFT FEASIBILITY

Field testing of various artificial lift methods to determine their applicability can be costly. Although plunger lift is a relatively inexpensive technique (approximately $4000 for a "minimum" installation), additional equipment options can increase the initial costs. Also, downtime for installation, adjustments to see if the plunger installation will perform, and adjustments to optimize production will all add to the costs.

To alleviate these costs, methods have been developed to predict whether plunger lift will work in advance of the installation, under particular well operational conditions. Although these methods vary in rigor and accuracy, they have historically proved useful when predicting the feasibility of the plunger lift method.

Several screening procedures can be used to determine if plunger lift will work for a particular set of well conditions.

7.4.1 GLR Rule of Thumb

The simplest of these is a simple rule of thumb that states that the well must have a gas/liquid ratio (GLR) of 400 scf/bbl for every 1000 ft of lift or some value that is fairly close to the 400 approximate value. (This corresponds to approximately 233 m^3 gas/m^3 liquid for every 1000 m depth.)

Example 7.1

Will plunger lift work for a 5000-ft well producing a GLR of 500 scf/bbl?

Applying the rule of thumb of 400 scf/bbl for each 1000 ft of lift, the required GLR is:

$$\text{GLR, required} = 400 \text{ scf}/(\text{bbl-1000 ft}) \times 5 = 2000 \text{ scf/bbl}$$

However, the actual producing GLR is 500 scf/bbl; so this well is not a candidate for plunger lift, according to this rule.

Although useful, this approximate method can give false indications when the well conditions are close to those predicted by the rule of thumb. Because of its simplicity, the simple rule method neglects several important considerations that can determine plunger lift's applicability. This rule of thumb, for instance, does not consider the reservoir pressure and resultant casing buildup operating pressure that can play a pivotal role in determining the feasibility of plunger lift. Well geometry (i.e., whether or not a packer is installed) can also determine if plunger lift is feasible In general, the 400 scf/bbl rule is for the case of no packer in the well.

7.4.2 Feasibility Charts

To alleviate some of the shortcomings of the GLR rule-of-thumb requirement, charts from Beeson et al.[2] have been developed that provide a more accurate way to determine the applicability of plunger lift. These are shown in Figures 7-6 and 7-7, which examine the feasibility of plunger lift for 2 3/8- and 2 7/8-inch tubing, respectively.

With reference to the charts, the horizontal X-axis lists the "net operating pressure." The net operating pressure is the difference in the operating casing buildup pressure and the separator or line pressure to which the well flows when opened.

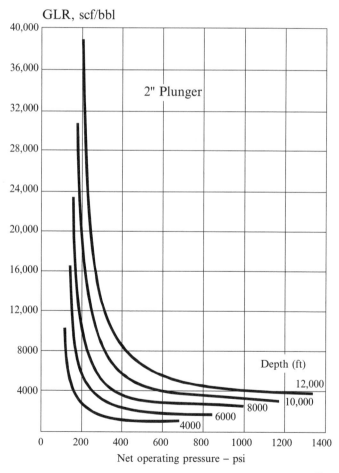

Figure 7-6. Feasibility of plunger lift for 2 3/8-inch tubing.[2]

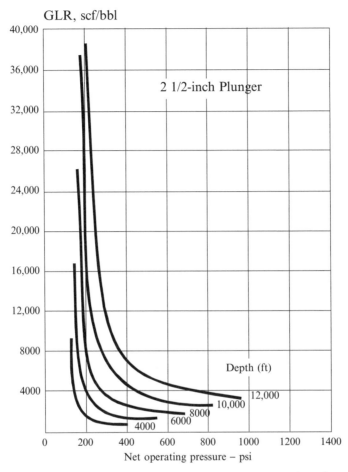

Figure 7-7. Feasibility of plunger lift for 2 7/8-inch tubing.[2]

The casing buildup pressure represents a casing pressure to which the well builds within a reasonable "operating" period. Because this time dictates the time permitted for each plunger cycle, reasonable time suggests a matter of a few hours rather than days or weeks. Although the line pressure used in the net operating pressure is more straightforward, some special considerations deserve mentioning. The line pressure used to enter the chart must be the flowing wellhead pressure. Often, if the separator is located a significant distance from the well, and particularly if the two are connected through a small diameter flowline, the line pressure might build when the well is allowed to flow. For example, if the separator pressure is 100 psi, the line pressure might build to 200 psi at the wellhead when the well comes on as the liquid slug is forced into the small diameter line. Therefore, the proper use of Figures 7-6 and 7-7 requires some judgment by the design engineer. The vertical Y-axis of the charts is simply the required minimum produced GLR in scf/bbl.

Use the figures by entering the X-axis with the net operating pressure. Track vertically upward to the intersection with the well depth. Then track horizontally to the Y-axis, and read the minimum produced GLR required to support plunger lift.

If the well's measured producing GLR is greater than or equal to that given by the chart, then plunger lift will likely work for the well. If the measured GLR of the well is close to the value given by the charts, the well may or may not be a candidate for plunger lift. Under these conditions, the accuracy of the charts requires that other methods be used to determine the applicability of plunger lift. The following example illustrates the use of the charts shown in Figures 7-6 and 7-7.

Example 7.2

A given well is equipped with 2 3/8-inch tubing (a 2-inch plunger approximately). Is this well a good candidate for plunger lift?

Operational data:

Casing buildup pressure	350 psi
Line or separator pressure	110 psi
Well GLR	8500 scf/bbl
Well depth	8000 ft

Use Figure 7-6 to determine whether plunger lift will work for this well.

Net operating pressure = (Casing buildup operating pressure −
Line pressure)
= 350–110
= 240 psi

Entering Figure 7-6 shows that at a depth of 8000 ft, the well is required to produce a GLR of approximately 8000 scf/bbl to maintain plunger lift.

The example well has a measured GLR of 8500 scf/bbl and is therefore a likely plunger lift candidate. Note that pressure, gas rate, and depth are accounted for in this chart.

A comparison between Figure 7-6 and Figure 7-7 suggests that there is an advantage to using the larger diameter tubing. As the tubing diameter increases, however, the likelihood increases that the plunger loses the liquid on the upstroke (because of liquid fallback around the plunger) and comes up dry. If the plunger comes up dry, the plunger (a large metal object) will impact the wellhead with great force, possibly causing damage. Because of this and other reasons, plunger lift is not as common with 3 1/2-inch tubing and, especially, with larger tubing sizes.

The casing size is not indicated in Figure 7-6 or Figure 7-7. Because the casing volume is used to store the pressured gas used to bring the plunger to the surface, the casing size is important. In general, the bigger the casing size, the smaller the required casing buildup pressure to lift the plunger and liquid. It is unclear if the figures were developed using 5 1/2-inch casing data, 7-inch casing data, or both.[2]

7.4.3 Maximum Liquid Production with Plunger Lift

Figure 7-8 helps evaluate the effect of liquid production rate on the feasibility of using plunger lift.[3] This figure shows the maximum possible liquid production rate that plunger lift will tolerate for a given depth and tubing size. The chart tubing size vs. depth in feet is on the X-axis and the maximum allowable liquid production in bbl/day for plunger lift on the Y-axis.

The chart is generally used by entering the X-axis with the well depth. Then, track vertically upward to the given tubing size. Finally, read horizontally to the left, and find the maximum allowable liquid production rate for the use of plunger lift on the Y-axis.

Example 7.3

A given well is 7000 ft deep and is to be produced by plunger lift through 2-inch tubing. What is the maximum liquid that can be produced?

Entering Figure 7-8 with the depth of 7000 ft and 2-inch tubing gives the maximum production by plunger lift of about 110 bbl/day. (The process is shown in the figure with the arrows.)

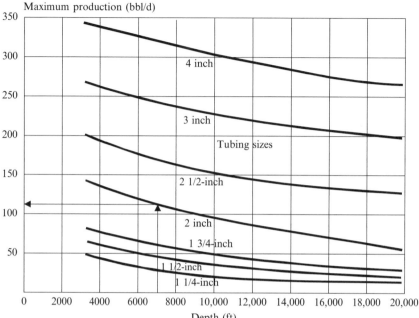

Figure 7-8. Liquid production estimate for plunger lift.[3]

7.4.4 Plunger Lift with Packer Installed

Although some installations have used plunger lift systems successfully in wells having packers installed, packerless completions are highly preferred. In the event that the well does have a packer installed, perforation of the tubing above and near the packer (thereby allowing the casing annulus to accommodate gas storage) can drastically improve the efficiency of the plunger lift system. However, packer liquid may have to be drained from the well annulus before going on production, perhaps by setting a plug below the perforated tubing section and bailing the liquid out of the well.

Some wells, however, have sufficient reservoir pressure and gas flow to produce liquids with plunger lift, even with a packer. When a packer is installed in the well, Figure 7-9 can be used to estimate whether the well conditions are sufficient to support a plunger lift system.

This figure plots two curves that represent the upper limit of conditions required for plunger lift for the cases with and without

Plunger Lift

Gas requirement for operating plunger pump

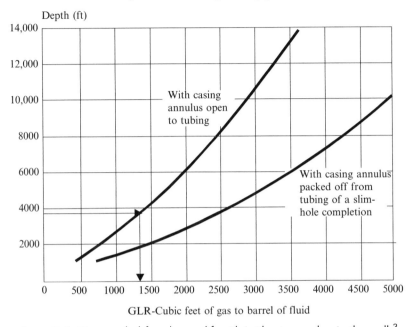

Figure 7-9. Gas needed for plunger lift with/without a packer in the well.[3]

packers installed in the completion. These are plotted against the GLR on the X-axis and the well depth in feet on the Y-axis. If the intersection of the well's GLR and depth falls on or below the respective curve, then plunger lift will likely work for the well. This figure clearly shows the adverse effects that packers can have on plunger lift installations.

For example, a well having a GLR or 1400 scf/bbl is sufficient to operate a plunger to a depth of 3900 feet if the well has no packer. With a packer installed, however, the operable depth is reduced significantly to 2000 feet. The general accuracy of this simple chart for wide application is suspect, but it shows the trends to be expected.

7.4.5 Plunger Lift Nodal Analysis

Reference 4 calculates the average bottomhole pressure for all portions of the cycle for one production rate. The average pressure includes the rise, the flow period, the fall period, and the buildup period. This is compared to various sizes of tubing and to what pressure is required to flow up the tubing at various rates. Then the plunger lift performance

Plunger vs velocity string performance

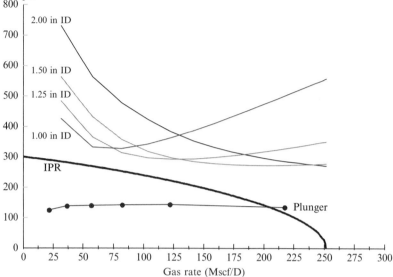

Figure 7-10. Showing plunger and smaller tubing performance on the same down-hole Nodal plot.[4]

can be compared to flowing up various sizes of tubing. The results of this type of analysis are shown in Figure 7-10.

Figure 7-10 shows the well inflow performance or IPR plot and the tubing performance of several sizes of tubing. For this example, none of the tubing performance curves are predicted to flow as they do not intersect the well inflow performance curve. However, the plunger performance[4] shows that for the low gas rates, using plunger gives a lower required flowing bottomhole pressure. As the well IPR declines to lower and lower pressure, only the plunger performance curve can intersect the inflow curve and achieve a flow rate.

7.5 PLUNGER SYSTEM LINE-OUT PROCEDURE

The following section outlines hints and suggestions to incorporate into the procedures used to bring a plunger lift system online. The section covers all aspects of plunger lift from the initial startup, considerations before and during the first kickoff of the plunger, methods to adjust the plunger cycle, and techniques to optimize the plunger cycle to maximize production. The following material on system operation and maintenance follows the Ferguson Beauregard Plunger Operation Handbook[5] (with permission) with some updates and alterations.

7.5.1 Considerations Before Kickoff

Several parameters must be considered before kicking off a plunger lift well. The most important is the casing pressure. As mentioned the casing annulus acts as energy storage, holding compressed produced gas that eventually is responsible for bringing the plunger and the liquids to the surface. It is this gas trapped in the casing that primarily determines the frequency of the cycles and therefore the success of a plunger lift system.

Another key factor to consider is the liquid load or the amount of liquid accumulated in both the casing and the tubing. The rate of accumulation of liquids also plays an important role in determining the plunger cycle time. If the liquid volume becomes too high, the plunger will be less likely to be able to bring the liquids to the surface using the gas pressure available.

A third major factor to be considered is backpressure. This includes backpressure from all likely sources—high-line pressure, small chokes, or compressors that will not handle the initial surge of gas. Backpressure is the pressure the well sees on the downstream of the tubing valve when it is opened.

Load factor

It is extremely important to properly prepare the well before you open it to flow. First, it should be as "clean," or as free of liquid, as possible. This may mean swabbing the well until it is ready to flow, or it may mean leaving it shut-in for several days to allow the well pressure to build and push liquids back into the formation.

The Load Factor can be used to see if the well is ready to be opened. The definition is as below:

$$\text{Load Factor} = 100 \times \frac{\text{Shut-in Casing Pressure} - \text{Shut-in Tubing Pressure}}{\text{Shut-in Casing Pressure} - \text{Line Pressure}}\%$$

A rule of thumb is to ensure that the load factor does not exceed 40% to 50% before opening the well to let the plunger and liquids rise.

Gas Well Deliquification

Example 7.4

A given well has been shut-in until the following conditions prevail. Determine whether the conditions are sufficient to start the plunger cycle.

Casing pressure:	600 psi
Tubing pressure:	500 psi
Sales line pressure:	100 psi

$$\text{Load Factor} = 100 \times \frac{600 - 500}{600 - 100}\% = 20\%$$

Because the Load Factor is less than the maximum limit of 50%, the plunger and liquid slug are predicted to rise when the well is opened.

Conditions are predicted to be acceptable to start the plunger cycle.

It is important to be patient while waiting for the well conditions to meet the initial load factor requirements. If the well is opened too soon without sufficient casing buildup pressure, or too large a liquid slug, the plunger may not make it to the surface, and the well will further load with liquids. The well should be allowed to build an abundance of pressure (more in fact than is actually needed) before opening the well to production. If time permits, the initial shut-in might be allowed to proceed until the pressures are static, just to ensure that this vital first cycle can be accomplished.

A common mistake is to allow the well to flow too long following the production of the initial slug after swabbing. Once the well's production becomes gaseous and the casing pressure begins to drop, the well should be shut-in and allowed to build pressure during the start-up period. The produced gas pressure is a vital component required to bring the plunger to surface and should be conserved, especially when just starting an installation.

In many cases, it is desirable to vent the gas above the liquid in the tubing on the initial cycle to a lower pressure. This creates more differential pressure across the slug and plunger, pushing the slug to surface. If this is not feasible, then every effort should be made to remove as many restrictions in the flowline as possible. If a flowline choke is required, the largest possible choke for the system should be used. It is also good practice to put large trims in the dump valves of a separator. A slug traveling at 1000 ft/min corresponds to a producing rate of 5760 bpd in 2 3/8-inch tubing. Frequently, a larger orifice plate in the sales meter is used to measure the peak flow of the head gas. Some controllers will perform the start-up techniques, but the details are presented for understanding.

7.5.2 Kickoff

Once adequate casing and tubing pressures have been reached, the well is ready to bring the plunger to the surface. The casing and tubing pressures required to kick off the well are obtained from the methods outlined previously.

It is imperative to open the motor valve as rapidly as possible so that the tubing pressure is bled off quickly. If done, this quickly establishes the maximum pressure difference across the plunger and the liquid slug to move them to the surface.

Record the time required for the plunger to reach the surface. The current thinking is that the plunger should travel at between 750 and 1000 ft/min average rise velocity for optimum efficiency. Experience has shown that plunger speeds in excess of 1000 ft/min tend to excessively wear the equipment and to waste energy and that lower plunger speeds will allow gas to slip past the plunger and liquid slug, lowering the system efficiency. The plunger travel speed is controlled by the casing buildup pressure and the size of the liquid slug that is produced with the plunger. A plunger could be run slower if it had a very good sealing mechanism.

When the motor valve is opened, a surge of high-pressure gas from the annulus will be produced into the tubing to lift the plunger and liquid. As the gas rate at the surface bleeds down, a slug of liquid will be produced, followed by the plunger. Often, some liquid will follow the plunger. In most cases, when just starting the plunger cycles, it is best not to let the well flow more than a couple of minutes after the plunger surfaces. If the well is allowed to flow for too long, the casing pressure will decrease below the recommended limit, allowing too much liquid to accumulate in the annulus before the next cycle. If the volume of liquids becomes excessive, the well will be incapable of completing the next cycle.

With the plunger at the surface initially, close the motor valve and allow the plunger to fall. Gas begins to pressurize the casing and tubing for the next cycle. The plunger must also be allowed to reach the bumper spring. Once the casing pressure has regained its initial value, the cycle can be set for automatic operation if a few of these manual cycles are used to start the plunger operation. Many newer controllers will perform the starting procedure without manual intervention.

7.5.3 Cycle Adjustment

Liquid loading can occur not only in the tubing but also in the reservoir immediately surrounding the wellbore. Liquid accumulation in the reservoir near the wellbore can reduce the reservoir's permeability. To partially compensate for this, it is recommended to run the plunger on a "conservative" cycle for the first several days. A conservative cycle implies that only small liquid slugs are allowed to accumulate in the wellbore and that the cycle is operated with high casing operating pressures.

When setting the operating cycle for a plunger lift installation, one proven method is to use a casing pressure sensor in combination with a plunger arrival sensor to shut-in the well. This method provides consistent shut-in casing pressures for each cycle, as long as the well is shut in immediately after the plunger arrives at the surface. In so doing, the method essentially minimizes the time required for the next cycle. If casing pressure above the line pressure is used as a control guide, it will prevent trying to bring the system on when line pressures have drastically increased from one cycle to the next.

In summary, the kickoff procedure is outlined below:

- Check (and record) both the casing and tubing pressures. Apply the rule of thumb shown in Example 7.4.
- Open the well and allow the head gas to bleed off quickly. Record the time required for the plunger to surface (plunger travel time).
- Once the plunger surfaces and production turns gassy, shut the well in and let the plunger fall back to the bottom.
- Leave the well shut-in until the casing pressure recovers to the pressure it had on the previous cycle. It is better to return to the casing pressure in excess of line pressure.
- Open the well and bring the plunger back to surface, and again record the plunger travel time. Shut the well in.
- If this cycle has been operated manually, then set the timer and sensors to the recorded travel time and pressures.
- If you have no casing pressure sensor or magnetic shut-off switch, then it is necessary to use time alone for the cycle control. Allow enough time for adequate casing buildup and enough flow time to get the plunger to the surface. A two-pen pressure recorder can be a valuable asset under these conditions. By monitoring the charts, you can quickly compare the recovery time of the casing and adjust the cycle accordingly.

• Whichever approach you use, once you see that the cycle is operating consistently, leave it alone, and allow the well to clean up for one or two days until the liquids in the reservoir wellbore area have been somewhat cleared.

Although many new controllers will take care of the above steps, the stops have been listed to show what physically should be considered to start a plunger installation and also for the situation when newer controllers are not being used.

7.5.4 Stabilization Period

Because the formation adjacent to the well tends to load with liquids while the wellbore itself loads up, it generally takes some time for the well to clean up. Depending on the reservoir pressure and permeability, this cleanup could be accomplished in a day, or it may take several weeks. Optimization procedures are easier to implement after the well has stabilized.

During the cleanup period, the plunger cycles should remain conservative. This implies longer shut-in cycles and shorter flow times than will be used after the well has had time to clean up. As the well produces liquids and stabilizes, the buildup casing pressures should rise and the rate of liquid production should decline. It is important to continue to keep plunger average rise velocity at approximately 750 ft/min. As the well stabilizes, the plunger travel time will initially decrease, then become stable, indicating that the well is sufficiently clean to begin optimization. Although the buildup pressures are changing such that they *can* become larger, production optimization dictates that the cycle times be adjusted for shorter cycles so that smaller operating casing pressures can be used.

7.5.5 Optimization

Once the well has stabilized, the plunger cycle is ready to be optimized. The optimization procedure varies somewhat, depending on whether the well is a gas or an oil well. The first step in either case is to determine the operating casing pressure. This is done by incrementally dropping the surface casing pressure, required just before each cycle, by 15 to 30 psi, then allowing the plunger to cycle four or five times before dropping the pressure again. At each incremental casing pressure, record plunger travel time to ensure that the plunger speed stays close (+/−) to an average speed of approximately 750 ft/min.

If the plunger speed drops below 750 ft/min, then slightly increase the casing operating pressure and record the plunger travel time for several more cycles until the plunger speed stabilizes at a value slightly above the minimum. If, on the other hand, the plunger speed is above approximately 1000 ft/min, allow the well to flow longer after the plunger surfaces to allow more liquid to feed into the wellbore during each cycle. Eventually, the swings between the high and low casing pressures will stabilize with the plunger travel times within the desired operating parameters, indicating that the well is again stable at the new casing operating pressure.

This discussion assumes that many of the adjustments are made manually to clarify how the well can be controlled. However, again, many of these operations are now taken care of by the newer computerized controllers. These manual steps are presented for understanding.

The next step is to adjust the time for the well to flow with the plunger surfaced. In this case, an oil well is generally much simpler to set than a gas well. Oil wells generally have much lower gas/liquid ratios (GLR, scf/bbl) and, therefore, have much less gas available to push liquids to the surface.

Oil well optimization

To fully optimize the flow time for an *oil well*, it is necessary to install a magnetic shutoff switch in the lubricator at the surface to shut the well in on plunger arrival. Any reliable arrival transducer would serve the purpose. The switch activates the motor valve, shutting the well in immediately on plunger arrival, which saves the needed tail gas for the next cycle. The plunger then starts its return to bottom with only a small hesitation at the surface, shortening the cycle time and increasing liquid production.

This prevents the well from depleting the vital gas supply stored in the casing. Depleting this stored gas would require longer shut-in periods to rebuild pressure and, in most cases, would lower the overall liquid production.

If the casing pressure remains too high after plunger arrival, rather than allowing the well to flow gas after the plunger has surfaced, the recommended practice is to lower the casing operating pressure. This generally prompts an increase in production because the pressure against the formation is reduced. This type of cycle is described in the paper by Foss and Gaul.[6] The authors have witnessed oil wells on plunger making as much as 300 bpd from approximately 4000 ft.

Gas well optimization

Optimizing the flow time for a *gas well* requires more effort, if done manually, because the time that the gas is allowed to flow after plunger arrival is considerably longer than that of an oil well. An older method of optimization is as follows: The flow time for a gas well is optimized by continually adding small increments to the amount of time allotted to gas flow while recording the plunger travel time. These small increments should be added over several days to allow the well to regain stability after each change. As the flow time increases, the plunger travel time will decrease. Once the plunger travel time drops to approximately 750 ft/min, the flow time used is considered optimized. However, now the velocities mentioned are achieved but attention is given to the average pressure on the formation during the cycle, and this is minimized by allowing only small liquid slugs to accumulate during the cycle.

Optimizing cycle time

The methods mentioned above to examine rise velocity only establish cycles, but they do not optimize production. For instance, a large slug can be brought into the well during the flow period, and then a large casing buildup pressure will allow the plunger and liquid to be lifted to the surface at 750 fpm. This would exert a high average pressure on the formation, and the production would be reduced.

It would be better if a small slug of liquid is accumulated in the tubing during a brief flow period, and then only a small casing operating build-up pressure would be required to lift the slug at an average rise velocity of about 750 fpm. This would result in a smaller average pressure on the formation, and the production would be higher (Figure 7-11).

7.5.6 Monitoring

Any changes to the conditions at the surface will have an impact on the operation of the plunger lift cycle. If the sales line pressure were to decrease because of a lower percentage of liquid in the flow, the optimum flow time would increase. On the other hand, if the sales line pressure were to increase, the flow time must be shortened. Similarly, if the orifice plate size or choke settings are changed, then the appropriate changes to the flow time must be made.

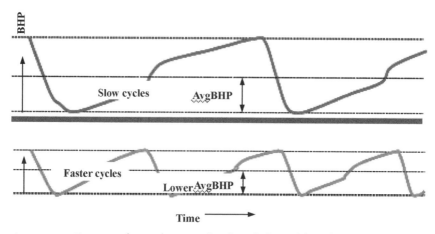

Figure 7-11. Faster cycles with a smaller liquid slug of liquid result in a lower average flowing well pressure.

Once the well is reasonably optimized and the plunger lift system is in stable operation, it is still necessary to monitor the well for best performance. Well and reservoir conditions continually change, thus altering the performance of the plunger lift system and requiring adjustments. Most controllers do this work, and the operator does not have to check on the performance. Controllers will also maintain the plunger rise velocity near 750 fpm or some input average rise velocity. It is a good idea to physically check the plunger for damage and wear monthly because plunger wear will also have an impact on the rise time for a given set of well conditions.

7.6 PROBLEM ANALYSIS

The following section outlines solutions to some of the more common problems encountered with plunger lift systems. These items are grouped with respect to the system components and particular malfunctions.

Table 7-1 (Phillips and Listiak[7]) summarizes various problems that can occur with plunger lift. It can be used as a quick reference. The following material revised from Ferguson and Beauregard[5] contains more detail of troubleshooting procedures.

Table 7–1
Plunger Lift Trouble Shooting Guide

Problem	Check/Change Plunger	Optimize Program Settings	More Off Time	More Afterflow	Less Off Time	Less Afterflow	Check Well TBG (Restriction/Hole)	Check Wellhead (Design)	Clean Sensor/Check Wiring	Check Module/Wiring	Change (+) Lead Fuse Link	Power Down Restart Module	Set Sensitivity of Sensor	Change Supply Gas Filter	Adjust Supply Gas Pressure (20–30psi)	Clean Control Bleed Ports	Change O-Rings Under latch Valve	Check/Change Battery	Check Solar Panel	Repair Motor Valve Trim	Eliminate Flow Restrictions	Check Catcher	Change Module	Change Latch Valve	Check Special Settings	Check Motor Valve Diaphragm	Inspect Plunger
No plunger arrival	6	3	2			1							4							9	10		8				
Slow plunger arrival	4	3	2			1	8	7												5	6	4					5
Fast plunger arrival	1	3		1	2		6	4					2									3					
Fast plunger arrival at all settings or plunger won't fall		1					4																				
Slow plunger arrival at all settings or plunger won't come to surface	4	3				1	7	6													5						
Short lubricator spring life	4	4	3	2			5	4														1					
Short plunger life	3	3	2	1			5																				
Sensor error	6	3							3	4			2									1	5				
Plunger error	6	3	2				12		5	7			4					9	9		10		8				
Good trip, no count (plug-in sensor)		1							5	3			2										6				
Good trip, no count (strap-on sensor)		1								1	4		2														
Fatal error code at LED										10	5	3													2		
LED control screen blank									5	5		4						2	3				5				
Sales valve will not open/close		1												4	3	6	7	2	8	9			11			13	
Tank valve will not open/close		1												4	3	6	7	2	8	9			10			12	
Latch valve will not switch														4	3	5	6	1	2	8			9	7			
Motor valves won't close or close slowly																1	2	5	6								
Short battery life		2																2	3				4				
Will not go to afterflow										3													4	1		1	

From Phillips and Listiak.[7]

7.6.1 Motor Valve

Valve leaks

When motor valves leak, there are two possible sources. Under normal conditions, a valve will have from 20 to 30 psi on the diaphragm section of the valve and much higher pressures on the body of the valve. External leaks are usually found at the packing section located between the diaphragm and the body of the valve. This occurs when the packing around the stem wears and leaks because of the high pressure from the body of the valve. All valves have some type of packing around the stem of the valve. In some cases, it is possible to tighten a packing nut and stop the leak; generally, however, it is necessary to replace the packing to eliminate the leak. Contact the valve manufacturer or the plunger lift company to help with the repair and parts.

The diaphragm portion of the valve can leak at one of two places: Either the valve will leak around the flange where the two portions of the diaphragm assembly are connected or at the breather hole (located on the opposite side from where the supply gas enters the diaphragm). In the latter case, the leak indicates a ruptured diaphragm. It is possible that a leak occurring at the flange can be the result of loose bolts, which can be corrected by simply tightening the bolts and nuts, eliminating the need to replace the diaphragm.

Internal leaks

The most common leaks encountered in motor valves are internal leaks. Often, ball and seat configurations are normally used as the sealing element. Because of the extreme pressure differential and high flow rates, the seat area is subject to fluid cut or erosion, which can be aggravated by abrasive materials. If the valve has an insert seat, it will have an O-ring seal, which is also susceptible to cutting or deterioration because of gas composition.

If a leaking valve is suspected, the leak can be isolated by simply putting pressure on the upstream side with the valve closed and checking to see if there is any flow through the valve. If flow is identified, the leak is likely across the seat and can be corrected in the following ways:

- Check the valve adjustment. Depending on the size of the seat, the size of the diaphragm, and the flow path, there is a maximum pressure that a particular valve will hold. Manufacturers have charts for determining this differential pressure.

- If the valve seat is subject to a higher pressure difference, it is possible that the diaphragm and spring cannot contain the pressure. If the valve is equipped with an adjustment bolt on top of the diaphragm, tightening down on this bolt will put more pressure on the ball and seat to seal against the higher pressure. Do not screw the bolt all the way in because it will restrict the valve from fully stroking open.

- Also, consider using a smaller seat. The differential pressure across the area of the seat prevents the seat from holding. A smaller seat can dramatically reduce the force against the diaphragm spring. If a smaller seat is objectionable, consider larger diaphragm housing. The larger housing will have a larger spring and can hold a higher differential pressure.

- The valve may be turned around in the flow. This will put the higher pressure on top of the seat and that pressure will help hold the valve closed. Caution should be exercised, however, because if the pressure is in fact too high for the particular seat, then it will prevent the valve from opening. This is a last resort before new equipment is installed, as this repair will make the valve chatter.

- Another cause of a leaking ball and seat can be the formation of hydrates (i.e., an "ice" composed of hydrocarbons and water) in the seat area. An extreme pressure drop across the ball and seat in some service will prompt the formation of hydrates. Correcting the leak under these conditions is accomplished by dissolving the ice or hydrate at the valve. With the hydrates removed, the valve should hold.

The prevention of ice or hydrate formation presents a somewhat more complex problem. The formation of hydrates might be prevented by either reducing the pressure differential across the valve or by increasing the temperature. Simply using a larger trim in the valve will not reduce the pressure drop. The best solution is to lower the operating pressure of the entire system. This is not always possible, however, because operating pressure directly affects plunger system efficiency.

A common but expensive method to solve hydrate problems is to inject methanol just upstream of the freezing point. Alternately, a choke (larger than the valve seat) can be placed downstream of the valve. This will reduce the pressure drop across the valve seat and can reduce or eliminate the formation of hydrates by spreading out the pressure drop. Also, a heat supply can be added.

Valve will not open

Four factors play a part in the opening or closing of a motor valve:

- Size of the diaphragm
- Amount of pressure applied to the diaphragm
- Compression of the diaphragm spring
- Line pressure acting with or against the valve trim

A malfunction of any one or a combination of these components can prevent the valve from properly opening.

It was pointed out before that too much line pressure acting on top of the trim of the valve could hold the valve closed. In this situation it is possible to increase the supply gas pressure to the diaphragm to help open the valve. Do not exceed approximately 30 psi diaphragm pressure when attempting this procedure. If the valve still won't open and the adjusting screw has been backed out, then change to the next smaller seat or use a larger diaphragm. Exceeding the 30 psi limit placed on the diaphragm gas pressure can cause the valve to bang open, which can cause damage or can rupture the diaphragm.

Another reason for a motor valve not opening is the adjustment of the compression bolt. The compression bolt puts tension on a closing spring that is connected to the trim by a short stem. If the compression bolt has been overtightened, the valve will not fully open. When flowing over the seat, the tension should be at a minimum.

Finally, if the above items have been checked and the valve still will not open, then the valve may have severe mechanical problems, such as a bent stem or a clogged valve. A bent stem or a frozen or clogged valve, although uncommon, is not out of the question.

Valve will not close

Many of the steps mentioned above are appropriate for troubleshooting a valve that will not close. In addition,

- Line pressure that is out of the operating range for the diaphragm size can prevent closure.
- The top adjusting bolt unscrewed too far could also prevent the valve from closing.

- Under certain conditions, it is common for ice to form in the trim, preventing the ball and seat from making a complete seal, thus keeping the valve open.
- Sand, paraffin, welding slag, or other foreign objects can get lodged between the ball and seat, preventing valve closure.
- If the controller is not allowing the supply gas to bleed, the valve will not close. If this problem is suspected, the compression nut on the copper tubing link to the motor valve should be loosened while operating the controller open and closed. This should free the controller to bleed the supply gas.

7.6.2 Controller

The most complex part of the plunger lift system is the controller. There are many commercial controllers, and description and analysis of each is beyond the scope of this text. The following discussion covers only those basic components that might apply to most controllers. Basically, all controllers have similar operational characteristics. Generally, they use a 20- to 30- psi pneumatic source, usually gas, which is used to open and close a motor valve. The motor valve is opened by directing supply gas through the controller to the valve diaphragm to force the valve open. The motor valve is closed when the controller blocks the supply gas and bleeds the gas from the diaphragm that opened the valve, thus allowing the valve to close. The discussion of controller troubleshooting will be in two sections: the electronics and the pneumatics.

Electronics

If the controller does not appear to be working properly and faulty electronic equipment is suspected, the first thing to check is the display. In addition to showing the time, most controllers are designed so that the display will indicate the mode of the controller (whether it is on or off), if it has power, or whether there are any outside switch contacts. No display may simply mean no power, so check the batteries for charge and proper contact.

If the batteries are OK, then check the controller mode to ensure that the unit is turned on or off. Check the motor valve and verify that it is in the corresponding position. If so, examine the gauges at the bottom of the controller. The gauge on the left reports the supply pressure coming into the controller while the gauge on the right reports the pressure in the line from the controller to the motor valve. The pressure supplied to the controller should be from 20 to 30 psi. If the pressure is below 20 psi, then the problem is with the supply pressure source.

Check the module by making sure that an electric pulse is produced by the controller each time it is switched from on to off. This discussion may or may not apply to some newer controllers.

Pneumatics

Controllers typically use some type of interface valve to control the pneumatic signal, and the two most common are primarily the latching valve and possibly the older slide valve. Each operates differently, but essentially performs the same function. The latching valve is composed of an electromagnet and a small poppet valve. The valve operates when an electric "on" pulse from the electronics module activates the magnet and pulls the poppet off its seat then latches it back, directing supply gas to the motor valve. The off pulse from the electronics reverses the polarity of the electromagnet releasing the poppet, and a spring moves it to the closed position. In the closed position, the poppet valve blocks the supply gas to the diaphragm and vents the gas, closing the motor valve. The slide valve consists of housing and a small piston that slides through a cylinder in the housing. The travel of the piston is limited by end plates. The piston is fitted with three O-rings; one at each end for power and one in the middle. The position of the piston, either at one end of the cylinder or the other, directs the gas or determines whether the valve is in the open or closed position. A solenoid is fixed to each end plate of the housing. When either solenoid receives an electronic signal from the controller, it directs a shot of gas to the power end of the shift piston, pushing it to the opposite end of the cylinder, thus opening or closing the valve. When the valve slides to the open position, supply gas is directed to the diaphragm of the motor valve; when it slides to the closed position, the supply gas is blocked, and the diaphragm is bled.

Troubleshooting and maintenance of these valves is performed in the same manner. If a pneumatic problem is suspected, the gauges on the bottom of the controller should first be analyzed. With supply gas being fed to the controller, when the controller is pulsed to "on," both gauges should read the same pressure. Then, if the controller is pulsed to "off," the pressure on the right-hand gauge should drop to zero. If this is not the case, then the likely problem is a faulty valve (shifter) in the controller.

The fact that the shifter (latching valve or slide valve) is not working does not necessarily mean that it is damaged. The shifter does require voltage. Once you have determined that the shifter is not operating, the next step is to check its supply voltage. Check the wiring to ensure that there are no loose connections or broken wires. Next, use a voltmeter to check that power is being supplied to the shifter from the electronics module. There should be no power supplied to the shifter until the controller is pulsed on or off when only a brief pulse is issued. If no pulse is evident, then the electronic module must be replaced. If the pulse is being fed to the shifter and it is not operating then the problem is with the shifter.

The most common problem encountered with the shifters is fouling from contaminated supply gas. Fortunately, shifters are easily disassembled and cleaned. After a thorough cleaning, the slide valve must be lubricated with a thin coat of lightweight grease (such as Parker O-ring Lube). It is not recommended to disassemble the solenoid valves located at either end plate of the sliding valve for cleaning. Although the solenoid valves rarely malfunction, when they do, they must be replaced.

To ensure smooth operation of either type shifter, a filter should be installed in the supply gas line to keep impurities from entering the shifter mechanism. The supply gas should also be kept as dry as possible. If casing head gas is to be used, it is good practice to install a drip pot upstream of the controller and keep it blown dry.

7.6.3 Arrival Transducer

The arrival transducer is a device that plays a very important role in most plunger lift installations. Its function is to detect the arrival of the plunger in the lubricator. This then typically signals the controller either to shut-in the well (oil well) or to switch valves or just to register the cycle in a plunger counter (gas well). Older switches use a magnet to close a set of contacts on an electric switch. This switch closure completes a circuit that sends a signal to the controller. Although these switches are normally trouble-free, mechanical malfunctions are possible.

Newer strap-on switches sense a field change in the tubing when the plunger arrives.

When the plunger is surfacing, but not going far enough into the lubricator to trip the switch, adjustments must be made to the system to let the plunger pass farther into the lubricator. To ensure that the plunger travels far enough into the lubricator to make contact with the sensor, the upper flow outlet should be open to allow flow to go past the sensor, carrying the plunger past the sensor and allowing the sensor to signal arrival of the plunger. Some plunger wellheads try to use only one outlet; however, the dual outlet is a better setup for the arrival transducer. As an alternative, the sensor is easily strapped near the bottom or below the lower outlet.

7.6.4 Wellhead Leaks

Wellhead leaks must be repaired to maintain a safe and clean environment at the well site. On most wellhead hookups, leaks are generally caused by faulty threads. Leaking around the wellhead bolts is typically caused by improper torque on the bolts, improperly repaired wellhead, or damaged bolts.

Other than the bolt connections, the most common place for wellhead leaks is at the catcher assembly or where the lubricator screws into the flow collar. The catchers usually are attached by some type of packing gland (not unlike those found on many valves). Leaks that occur at the catcher can normally be fixed by tightening the packing nut. If not, it may be necessary to replace the catcher assembly.

The lubricator upper section has a quick connect that has an O-ring seal. These can leak and need to be replaced periodically. In most cases, tightening will not stop a leaking O-ring.

7.6.5 Catcher Not Functioning

For plunger inspection but not general operation, the catcher should be able to hold the tool in the lubricator to accommodate its removal. Plunger catchers catch or trap the tool and hold it:

• Catchers use a spring-loaded cam-type device. To activate the catcher and catch the plunger, either a thumbscrew is unscrewed (which activates the catcher) or a catcher handle is released. In both cases, a cam is extended into the path of travel of the plunger. As the plunger moves past, the cam is pushed back, allowing the plunger to move past it. Once the plunger has moved past the catcher, the spring-loaded cam flips out beneath the plunger, preventing it from falling back downhole.

- The other type catcher, which is commonly found on older installations, uses a friction catch to hold the plunger at the surface. The friction-type catcher consists of a ball extending into the sidewall of the catcher, pushed by a coil spring. As the plunger moves past the ball, the compression of the spring on the ball causes friction against the side of the tool, preventing it from falling.

Before troubleshooting catcher problems, first verify that the plunger is arriving at the surface and then make sure that on arrival the plunger is traveling far enough into the lubricator for the catcher to engage. One way to help the plunger go farther into the lubricator is to open the flow outlet above the catcher. Also, closing the lower outlet will direct all the flow through the upper outlet, driving the tool higher into the lubricator. If, under these conditions, the catcher still fails to capture the plunger, then further inspection of the catcher is required.

The first step in troubleshooting the catcher is to see if ice, paraffin, or other produced solids have clogged the catcher. The removal of these foreign materials should restore catcher operation.

Next, inspect the catcher nipple while manually engaging and disengaging the catcher. The nipple should move all the way out of sight and stay there. In the run position when the catcher is activated, the cam (or ball) should extend into the path of the plunger. If it is not extending or if it is retracting back into the housing, then it needs to be repaired or replaced. Never operate with the plunger surfacing and the wellhead open at the surface.

7.6.6 Pressure Sensor Not Functioning

An older but still used method for starting many plunger cycles is with a casing pressure-activated switch-gauge. The switch-gauge is a pressure gauge with two adjustable contacts on the face and a pressure indicator (needle); all of which are connected to electric wires. Changes in casing pressure (up or down) can cause the needle to move toward one contact or the other. When the needle touches either the high or low contact, it completes a circuit, which signals the controller to open or close the motor valve. Although rare, switch-gauge malfunctions do occur. Malfunctions include poor electrical contacts & pressure line problems.

Newer controls are pressure transducers that feed information to electronic computerized controllers that provide control logic from various algorithms.

7.6.7 Control Gas to Stay on Measurement Chart

Controllers may be used to throttle motor valves open or closed while maintaining a set-sensed pressure. When used in conjunction with a plunger lift system, its purpose is to restrict the initial surge of head gas within the pressure limits of the system to prevent the produced gas from going off the sales chart. Although these controllers are commonly used on compressors and production units, they are finding application with plunger systems. It would be better to have an electronic sensor that will record the bursts of gas, however, because throttling back the surge of head gas can only serve to have some effect in *reducing the production.*

The unit works by sensing a pressure and then converting that signal to a proportional pneumatic pressure to the diaphragm of a motor valve, causing the motor valve to throttle. The sensed pressure pushes on a high-pressure flexible element, which in turn operates a pilot valve. This throttling of the pilot valve varies the pressure supplied to the motor valve causing the motor valve to respond in a manner directly proportional to the sensed pressure signal. By throttling the motor valve, the unit attempts to maintain a constant sensed pressure. If the system (well) cannot supply enough pressure to meet the throttling range preset, however, the motor valve will remain wide open. On the other hand, if the sensed pressure exceeds the preset design pressure maximum, the motor valve will close completely.

This system has have two weaknesses. First, the supply gas entering the controller is metered through a small choke or orifice. When the gas supply leading to the output signal is slow to respond (build), this orifice should be examined. The choke is very small and is prone to clog with debris from dirty supply gas. It can usually be cleared with a small wire. It is good practice to place a filter in the supply gas line upstream of the choke to help prevent this clogging. If the controller does not respond to sensed pressure, the sensing element should be inspected.

7.6.8 Plunger Operations

Plunger will not fall

Plungers are free-traveling pistons that depend solely on gravity to get back to the bottom of the well. If the plunger remains in the wellhead after the shut-in period or if it is back at the surface very quickly after opening the well, there is likely an obstruction either in the lubricator or downhole keeping the plunger from falling to bottom.

In the event that the plunger returns to surface too quickly, first make sure that the plunger has been given ample time to reach bottom. Ideally, a plunger should travel up the hole between 750 and 1000 ft/min. ft/min. On the other hand, plunger fall rates can be considerably slower. To allow gas to easily flow through the plunger on the down cycle, plungers without a bypass can fall at rates of only 250 to 500 ft/min or greater. Plungers equipped with a bypass or collapsible seal may fall at rates between 500 and 1000 ft/min. Fast fall is recommended to optimize a system for high production. If liquids have accumulated in the well during the last bit of afterflow, then for maximum production the well should be opened as soon as the plunger lands on the bumper spring.

Echometer, Inc.[8] has devised a system that tracks the plunger during both the rise and fall portions of the cycle. The system is not commercial for individual wells but can be used on a well for analysis and removed afterwards. The measurements have been made both by acoustically recording the plunger depth using acoustic pulses generated with a gas gun and also by the pressure change that occurs as the plunger travels past the tubing joints. Figure 7-12 is a schematic of the Echometer setup to record plunger travel with time. Figure 7-13 is an example of the pressure and acoustic trace of a plunger cycle.

If the plunger ran smoothly during the initial installation, then it is unlikely that tubing is either crimped or mashed. If damaged tubing is suspected, then a wireline gauge ring should be run in the tubing having an OD corresponding to the tubing's manufactured drift diameter. It is also a good practice to run a gauge ring with a gauge length at least the length of the plunger. Care should be exercised when running the gauge ring, however, to prevent the ring from becoming stuck in the event that foreign debris is in the tubing. See Figure 7-14.

If the tubing, is assumed to be of good quality, then the two most common ailments that prevent the plunger from reaching bottom are ice (hydrates) or wax (paraffin) deposits. Typically, plungers will scrape the tubing clean of paraffin when cycling frequently. Severe paraffin buildup generally must be cut out of the tubing with a wireline cutter.

Hydrate formation often occurs in particular gas wells at a depth (approximately 3000 ft) where the gas is expanding rapidly. If the well is plagued with severe hydrate problems, a methanol injection system may be required to bring the well back to normal operation.

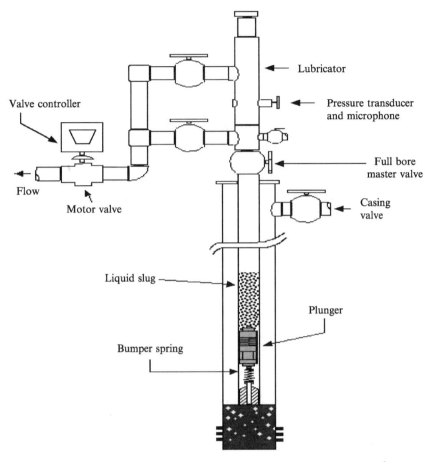

Figure 7-12. Echometer well configuration for plunger lift analysis.[8]

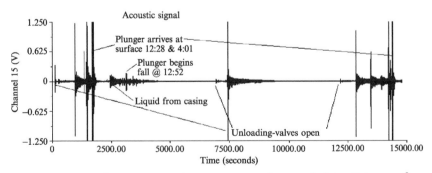

Figure 7-13. Sample acoustic and pressure signals recorded by Echometer[8] to monitor plunger travel.

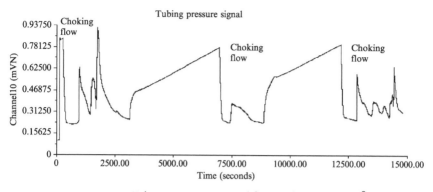

Figure 7-14. Tubing pressure record from Echometer tests.[8]

If the plunger will not drop out of the lubricator at the surface, the most likely cause is a faulty or damaged catcher. Review the preceding section on catchers.

Finally, if the well was recently worked over or other malfunctions have occurred, then there might be foreign debris (e.g., catcher parts, old swab cups, sand plugs) lodged in the tubing, which are preventing the free travel of the plunger. Run a wireline gauge ring. In addition to foreign material in the tubing, the plunger may have damaged or bent parts, which are impeding travel. Check that the pads on the plunger move freely. Sand behind the pads makes the tool stiff and difficult to fall, and a brush plunger could be required for sandy production.

The Echometer[8] portable system can be used for on-site analysis and can reveal many of the potential problems discussed above.

Plunger will not surface

Plunger lift operations require the tool to travel the full distance between the bottomhole spring and the lubricator during each cycle. If the tool is not getting to the surface, some or all of the liquid load will remain in the well. It can be difficult to isolate the source of the problem preventing the plunger from surfacing. There are both mechanical and operational considerations.

The ideal travel time for a surfacing plunger ranges from 750 to 1000 ft/min. This is the ideal rate, however, and many installations operate at much slower speeds. It is important, therefore, that ample time be given for the plunger to travel to the surface. If the plunger has been given sufficient time to surface (corresponding perhaps to an equivalent 100–200 ft/min rise time), then other problems must be investigated.

First, inspect the system for mechanical malfunctions. Most of the mechanical problems that prevent a plunger from falling to bottom would also prevent it from rising to the surface. Debris in the tubing, tubing quality, and plunger damage can all prevent the plunger from reaching the surface. In addition to restrictions, however, conditions that prevent the plunger from sealing in the tubing can prevent it from reaching the surface. These would include ballooned tubing, mixed tubing strings with changes in the ID, tubing leaks, and gas lift mandrels installed in the tubing. Typically, when the plunger encounters enlargements and loses its seal, it will stop traveling at that point. It is vital that well completion records be checked closely before installing a plunger lift system.

Finally, the plunger itself may have been damaged, preventing it from surfacing. Plungers equipped with bypasses may develop leaks, which prevent an adequate pressure seal across the plunger. The plunger should be checked regularly for wear and loose parts. Although uncommon, plungers can come apart in the hole. In some cases where the plunger will not surface under normal conditions, it may be possible to bring the plunger up by venting the head gas to a low-pressure separator. This provides extra pressure differential across the plunger that may be sufficient to bring the plunger to the surface. If this fails, the plunger must be wirelined out of the well.

Operational problems that would prevent the plunger from surfacing have all been discussed in previous sections. It may be necessary to go through the initial kickoff procedure again to ensure that the well is ready to begin normal plunger operations. It is important to make sure that the casing is allowed to reach the required operating pressure. It might be necessary to allow the casing to come to equilibrium before attempting another plunger cycle. If the plunger has been idle for a time, it may be necessary to swab the well and produce most of the liquids or shut the well in for a period to drive liquids in the formation before attempting to start the plunger cycle.

Plunger travels too slow

The speed that the plunger travels to the surface can greatly affect the performance of the plunger lift system. Plunger travel speeds that fall below the suggested 750 ft/min can significantly reduce the efficiency of transporting the liquids. For high-rate gas wells, this may not be a critical problem, because they generally have ample gas production to replace that lost in inefficiencies. In weak or marginal gas wells, where gas production is low and all the available casing gas is needed to surface the plunger, this can be a very important issue.

The plunger and liquid slug rise with the aid of the gas stored in the casing annulus with some help from formation production as well. If there is not a large volume of casing gas available and/or if it takes long shut-in times to rebuild casing pressure, the maximum possible number of plunger cycles per day is less. Experience has shown that the slower the plunger travels, the less efficient it becomes and the more gas it takes to move it to surface as gas leaks past the plunger. The seal between the plunger and the tubing is such that some gas always slips past the plunger, reducing its effectiveness, because the pressure below the plunger is larger than above the plunger. When the plunger is traveling within the optimal speed range (750 to 1000 ft/min), this gas slippage is presumed minimal. As the travel speed falls below the optimum, however, the amount of gas slippage is increased. This means that more of the casing gas is used each cycle, and the shut-in (or buildup) time is longer. Ultimately, this results in fewer cycles per day, which generally amounts to less liquid production per day. It is important to maintain plunger speeds near the optimum so as not to waste valuable casing gas, particularly on low-rate wells. Many of these guidelines are from experience and may not have been tested extensively; therefore, questioning and testing standard procedures for your particular wells is a good idea.

A number of ways exist to increase the plunger rise speed while conserving casing gas and maintaining adequate liquid production. The plunger travel speed is a function of the size of the liquid load and the amount of net casing pressure (less sales line pressure), plus the rate that the head-gas is removed at the surface. As mentioned, the plunger speed also has a major effect on the efficiency of the plunger seals. It is important to first analyze the well conditions and determine whether smaller slug size or a higher casing operating pressure is warranted. By raising the casing operating pressure, more effective pressure is exerted against the formation, thus lowering the liquid influx and reducing the slug size.

Reducing the size of the liquid load then allows higher plunger rise speeds. Similarly, lengthening the shut-in time again raises the net pressure on the formation while increasing the amount of gas stored in the casing annulus. With roughly the same liquid load (or possibly less) but more compressed gas in the annulus at a higher pressure (more energy), the speed of the plunger is again increased.

On the other hand, reducing the sales line pressure has the same effect as increasing the casing pressure on the plunger travel time (more differential pressure across the plunger) without the adverse effects of longer shut-in periods and more pressure against the formation. Plungers with more efficient seals can also operate with reduced plunger travel times by reducing the amount of gas slippage. Often, just replacing plungers with worn seals will have a dramatic impact on performance.

Finally, plunger performance can be improved by rapid evacuation of the head gas (the reverse of trying to choke back gas surges to keep them on a recording chart) above the liquid slug. This could require replacing an existing orifice plate with a larger size, opening up a choke, or enlarging the dump-valve trim to allow greater use of the gas that is available.

Velocity controllers control the flow time and the buildup time to maintain the correct rise average velocity but do not necessarily tend to low operating pressures and shorter cycles needed for production optimization.

Plunger travels too fast

A plunger traveling up the well too rapidly, especially with no liquid load, could cause damage. Although the efficiency of the plunger sealing mechanism is not dramatically affected by higher speeds, well safety and equipment longevity dictate that the plunger rise speed be maintained below the 1000 ft/min maximum. The plunger and lubricator undergo fairly severe punishment under normal operating conditions. As the plunger speed increases, the impact force imposed on the lubricator by the plunger increases roughly by the square of the speed. Although the plunger and lubricator are designed to withstand plunger impacts under normal speeds, higher speeds can quickly wear out and destroy both. In general, the economic benefits brought about by longer equipment life far outweigh those of shorter plunger travel times. A large plunger coming up dry, such as for 2 7/8- or 3 1/2-inch tubing, can cause the most damage.

From an operational standpoint, either decreasing the casing buildup pressure or increasing the size of the liquid slug can reduce plunger travel speed. This can be accomplished by allowing the well to flow for longer periods after plunger arrival at the surface. Another way to accomplish this is to reduce the shut-in period. Although not recommended choking the well, to slow the plunger will sometimes accomplish the objective. Choking or operating with too large of a liquid slug reduces production.

One reason that a plunger is coming up too fast is that even though there was liquid in the tubing when the plunger fell the liquid can be displaced from over the plunger to the casing during the shut-in period. This could be caused by bubbles entering the tubing during shut-in or perhaps the casing liquid is dropped below the tubing end, which might accelerate the loss of liquid from over to under the plunger. Also, the spring can be set to automatically open if large slugs appear in the tubing.

One method to control this is to run a standing valve below the bumper spring. However, a standing valve would trap any random slug that might be too large to lift, and you could not then raise the tubing pressure to push the slug below the plunger to start the cycle again. The standing valve would hold the large slug over the plunger regardless of pressure changes.

One common method used to attack this problem is to use a standing valve, but notch the seat of the valve so it will leak. It will then give some resistance to liquids leaking back to below the plunger during the buildup cycle and liquids can still be forced below the plunger through the leak if the slug should be too large to continue the cycles.

Another method is to use a new spring-loaded seat on the standing valve (Figure 7-15). The standing valve holds liquids over the plunger during the off cycle; however, if the need arises to add tubing pressure to pressure liquids back below the plunger, then enough pressure can be applied to force the seat down and allow the liquid to leak back from over to under the plunger. Also, the spring can be set to automatically open if large slugs appear in the tubling.

7.6.9 Head Gas Bleeding Off Too Slowly

Bleeding the head gas off too slowly can reduce the differential pressure needed to surface the plunger. The slower the bleed, the less the differential pressure across the tool and the less the chance of the plunger surfacing. The faster the head gas is allowed to bleed, the better the plunger performance.

Small chokes and high flowline pressures act as large barriers for the head gas to overcome, thus keeping the system from performing at optimum efficiency. It is critical to get rid of the head gas as quickly as possible. If it is necessary to choke the well, the choke should be as large as possible. To accomplish this it may be necessary to modify the surface facilities; however, the benefit of doing so far outweighs the cost. If the sales line pressure is too high, then efforts should be directed toward

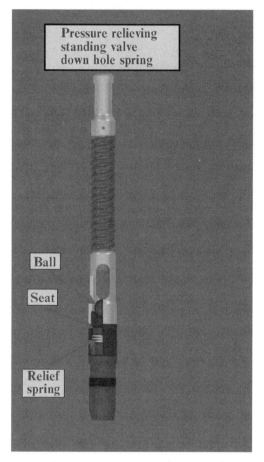

Figure 7-15. Schematic of a spring-loaded seat for a standing valve to be placed below the bumper spring (Ferguson Beauregard, Tyler, TX).

reducing that pressure, although this is a potentially expensive process that may require compression.

7.6.10 Head Gas Creating Surface Equipment Problems

A common complaint about intermittent operations is that they create problems with the surface equipment and gas measurement. Plunger lift falls in this category.

When a well fitted with plunger lift is first opened, a surge of high-pressure gas is generally forced at high rates through the surface equipment.

Often, the surface equipment was designed for an average flow rate and cannot handle the short duration surge that ends up going off the charts. One common but not recommended way of handling this problem is to install a positive choke in the flowline. Although the choke will restrict the initial gas surge to manageable levels, it will also restrict the flow of the remainder of the gas and the liquid slug. In particular, when a liquid slug passes through a gas choke the flow is drastically reduced, presenting a wall of liquid to the plunger that has the same effect as closing a valve. The consequence of this is a loss of production.

Fortunately, the problem is often negated once the well has been optimized. If this is not the case, however, other methods can be used. One of the most effective ways to correct the problem is to install a valve with a throttling controller (discussed previously) to limit downstream pressure while allowing the motor valve to be opened slowly to minimize production loss. This type of controller can be optimized to adjust to the pressure capabilities of the surface system and can therefore eliminate problems, such as selling off the chart, overpressuring separators, and surging compressors. Finally, in installations where several wells are on plunger lift, the surge effects can be negated by producing several wells through a manifold. In this manner, the surges produced by various wells can be timed to occur at different intervals, and any single surge will comprise a smaller percentage of the total flow and therefore be less likely to peg the sales meter.

7.6.11 Low Production

Optimizing or fine-tuning a plunger lift well can make a difference in the production. Consider testing short flow times to bring in small slugs of liquid. Then, short buildup times required to build smaller casing pressures are required to lift smaller liquid slugs. The result is a lower average flowing bottomhole pressure and more production. Limits are that a too short flow period could result in little or no liquid slug and a too short shut-in period would not allow the plunger to reach bottomhole. Also, related to the previous discussion, if the liquid is lost (all or some) during the shut-in period, low production would result.

7.6.12 Well Loads Up Frequently

Many wells are found to be very temperamental, where any small change in the operation can greatly affect their performance. Marginal wells tend to be particularly sensitive and are often easily loaded up.

Liquid loading on a plunger lift well is usually a result of too long of a flow time or too little casing pressure during the shut-in period. Also, trying to run plunger lift in small tubing can aggravate this problem.

A more conservative plunger cycle can alleviate liquid loading of a plunger lift well. As stated above, this means higher casing operating pressures and longer shut-in periods. Once the cycle has been changed, the well should be allowed to stabilize, which may take several days. Then continue with the optimization procedures outlined previously, making only small incremental changes to the system times and pressures, then allowing the well to achieve stability between each change. It is possible to eventually adjust the well back to the original cycle settings once it has had a chance to clean up.

If a well is completely loaded with liquids, then it must be brought through the kickoff procedures from the beginning. First, shut the well in and allow it to build pressure. With the well loaded, it may be necessary to swab the well to clean it up before starting the kickoff. For more permeable wells, a shut-in period will drive liquids into the formation. Remember to work slowly, making small incremental changes to the system and then allowing the system to become stable before continuing to the next step. Many new controllers now adjust cycle times and pressures to follow optimization algorithms.

7.7 NEW PLUNGER CONCEPT

A new two-piece plunger (MGM Well Service, Corpus Christi, TX; Figure 7-16) is designed to trip to bottom while the well is producing at considerable rate. In some wells, the plunger falls to the bottom while the well is producing at 1000 Mscf/D or more. Both pieces of the two-piece plunger have considerable bypass area when they are falling independently in the well, allowing the well to produce around the bottom piece (the ball) and through the top piece (the piston). They join at the bottom and are held together by the flow from the zones below as it pushes the plunger (now one unit) and any liquid in the tubing to the surface as would a conventional plunger system. The surfacing plunger strikes a shifting rod and a gas-powered catch cylinder. The shifting rod separates the two pieces, and the piston is held at the surface by the catch cylinder or in some cases by just the flow around the cylinder. The ball falls back to the bottom to await the arrival of the piston. When released from the surface, the piston arrives at the bottom of the well and joins with the ball, beginning the process again. The cylinder can be released by a short

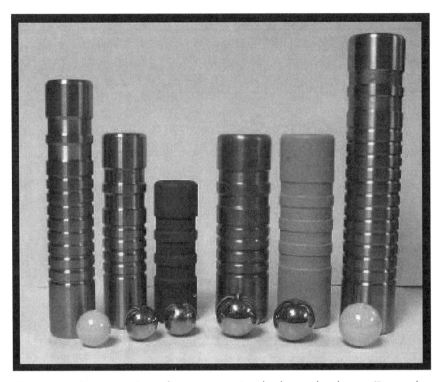

Figure 7-16. New two-piece plunger concept with plunger hardware. (Pacemaker Plunger; a division of MGM Well Service, Corpus Christi, TX.)

shut-in time so that pressure and fluid drag will cease holding the plunger at the surface. If the arrangement is not such that pressure and drag are holding the plunger cylinder at the surface, then a mechanical catch system may be used.

The plunger can trip to the bottom at speeds of 1000 ft/min or faster, while the well is flowing at a considerable rate. The high roundtrip speed allows the plunger to lift smaller amounts of liquid with each trip so it can make more liquid per day with less average bottomhole pressure than conventional plunger lift systems.

Another advantage of the two-piece plunger system is that it is stated to perform very well without using the casing or tubing annular volume for pressure storage. The plunger is stated to rely more on volume than trapped pressure to move the plunger to the surface. The two-piece plunger works in 2 7/8-inch slim hole or wells with a packer and no communication with the annulus. Wells with on-site compressors are usually adapted to the plunger because the "shut-in" time of only seconds has almost no effect on the suction pressure of the compressor. The smaller liquid loads have less effect on suction pressure, and the compressor may not need a recirculating valve. A shut-in time of only seconds does not create high spikes in wellhead pressures or volumes. The effect of the two-piece plunger is similar to a normal flowing well so it should not be necessary to "oversize" the compressor to accommodate the volume spikes common with the conventional plunger systems.

Because much of the operational practice and some of the feasibility charts in this chapter for conventional plunger systems consider the energy stored in the casing before opening the well to allow the plunger to rise, then these practices and charts should not be applied to use of the two-piece plunger. In general, this new concept looks interesting by eliminating or greatly reducing the buildup pressure time in the plunger cycle. Although this is new, the reduced or eliminated casing pressure buildup period has allowed increased production in some wells.

7.8 OPERATION WITH WEAK WELLS

Two methods are mentioned here using plunger lift for weak wells. One is the use of the casing plunger and the other is the use of a side string for gas injection.

7.8.1 Casing Plunger for Weak Wells

The casing plunger travels in the casing only, and there is no tubing in the well. The plunger senses when a head of liquid appears above the plunger, the internal bypass valve then closes, and the well gas production lifts the plunger and slug of liquid to the surface. Then at the surface, the plunger internal valve will open and the plunger will drop. The casing plunger rises and falls slowly. It has rubber cups that fit the casing. The cups will not last long enough for satisfactory service if the casing is very rough. Figure 7-17 shows a casing plunger.

Figures 7-18, 7-19, 7-20, and Table 7-2 are introduced to allow the sample calculation that follows to size cycle gas requirements for using a casing plunger. The calculations are from FBI (Tyler, TX) who no longer supply casing plungers. MULTI (Millersburg, OH) has been a supplier of casing plungers.

Figure 7-17. Casing plunger showing the rubber cups that seal against the casing ID and the lubricator and downhole casing stand (Multi Products Co, Millersburg, OH).

Application example for casing plunger:

FL pressure: 30 psi
Casing size: 4 1/2 × 11.6/ft
Depth: 50,000 ft
Production: 6 bopd, 4 bwpd, 40 Mscf/D
Surface shut-in pressure: 260 Psi
Combined specific gravity: 960

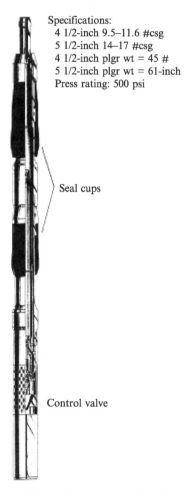

Specifications:
4 1/2-inch 9.5–11.6 #csg
5 1/2-inch 14–17 #csg
4 1/2-inch plgr wt = 45 #
5 1/2-inch plgr wt = 61-inch
Press rating: 500 psi

Seal cups

Control valve

Figure 7-18. Casing plunger internals (FBI, Tyler, TX).

Quick Look Calculations:

Total production is 10 bfpd. From Figure 7-19 for casing plungers, assume 2 bbl/cycle which will generate 5 cycles/day.

From Table 7-2, the minimum gas required is found for 30 psi line pressure and 2 bbl/cycle to be 705 scf/1000 ft/cycle or 0.705 scf/ft/cycle.

Depth is 5000 ft

Minimum gas required = 5 × 0.705 × 5000 = 17,625 scf/D

Actual gas required = 17.625 × 2 = 35,250 scf/D

Casing plunger production cycles

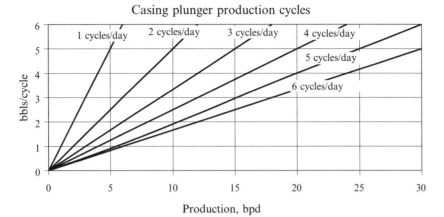

Figure 7-19. Casing plunger cycle design chart (FBI, Tyler, TX).

Casing plunger 4 1/2-inch × 11.6 #/ft

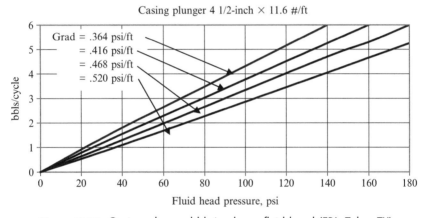

Figure 7-20. Casing plunger bbls/cycle vs. fluid head (FBI, Tyler, TX).

To correct for specific gravity:

Enter Figure 7-20 from left axis at 2 bbls/cycle, intersect the 0.96 SG line, and read down to find head pressure = 54 psi.
Casing volume = 0.0873 scf/ft
Friction – Weight Constant = 26 psi for 4 1/2 inch
Gas volume = (Pline + Pliquid + 26)/14.7 × 1000 × 0.0873
= (30 + 54 + 26)/14.7 × 1000 × 0.0873
= 653 scf/1000 ft/cycle or 0.653 scf/ft/cycle
Min gas required = 0.653 × 5 × 5000 = 16,325 scf/D
Actual gas required = 16,325 × 2 = 32,650 scf/D

Table 7-2
4 1/2 Inch Casing Plunger Gas Requirement, Minimum scf/1000 ft/cycle
(FBI, Tyler, TX)

				<	Line	Press	psi	>		
Bbl/cycle	10	20	30	40	50	60	70	80	90	100
1	400	460	520	580	640	700	760	820	880	940
2	585	645	705	765	825	885	945	1005	1065	1125
3	770	830	890	950	1010	1070	1130	1190	1250	1310
4	955	1015	1075	1135	1195	1255	1315	1375	1435	1495
5	1140	1200	1260	1320	1380	1440	1500	1560	1620	1680
6	1325	1385	1445	1505	1565	1625	1685	1745	1805	1885
7	1510	1570	1630	1690	1750	1810	1870	1930	1990	2050
8	1695	1755	1815	1875	1935	1995	2055	2115	2175	2235
9	1880	1940	2000	2060	2120	2180	2240	2300	2360	2420
10	2065	2125	2185	2245	2305	2365	2425	2485	2545	2605

In summary, the casing pressure can be used for shallow low-pressure wells with relatively good casing condition to extend the life of the well to very low pressures. The casing plunger can also be made for 5 1/2-inch casing. Multi-Systems can supply a casing plunger.

7.8.2 Plunger with Side String: Low Pressure Well Production

Plunger lift with a side string can be used to produce gas or oil wells with low bottomhole pressures where a source of higher pressure makeup gas is available at the wellhead. A plunger lift system in combination with a side string for injecting makeup gas and pressure for lift is used for this system.

The plunger lift system with side string injection requires that the tubing be removed from the well. As the tubing is run back in the hole, 1/2" or 3/4" coiled stainless steel tubing is banded to the production tubing. On bottom is a standing valve with a side port injection mandrel above it. Above the injection mandrel is a bottomhole spring assembly and a plunger.

Makeup gas is injected from the surface down the side string directly into the production tubing. As the gas enters the tubing, it is prevented from entering the wellbore by the standing valve. The gas is forced to U-Tube up the production tubing, driving the plunger ahead of it, which in turn removes liquid from the tubing.

The injection gas is only injected long enough to cause the plunger to surface. Once the plunger surfaces, the well is allowed to bleed down to sales line pressure. As this occurs, liquid enters the production tubing from the wellbore, and the plunger drops back to bottom on its own weight.

As the plunger continues to remove liquid from the wellbore, the liquid level in the casing drops. As the liquid in the casing drops, the perforation zone is relieved of hydrostatic pressure, and formation gas enters the casing. The formation gas is produced out the casing.

PLSI, Midland, TX, developed this technique for low bottomhole pressure wells in 1992 (Figure 7-21). Initial installations occurred in the Antrium gas zones of northern Michigan. The technology has been

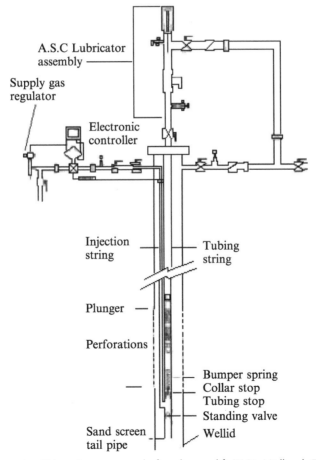

Figure 7-21. Side string gas supply for plunger lift (PLSI, Midland, Texas).

economic in this area and to date more than 500 installations of this system are in place.

If you have a compressor or a source of higher pressure gas, you can use this concept with the side string to lift liquids from low-pressure gas wells where it is not feasible to run a conventional plunger system.

7.9 PLUNGER SUMMARY

- Plunger systems work well for gas wells with liquid-loading problems as long as the well has sufficient GLR and pressure to lift the plunger and liquid slugs.
- Plunger lift works well with larger tubing so there is no need to down-size the tubing.
- Conventional plunger lift works much better if there is no packer, and this can be a problem if the old packer is removed.
- Although the plunger lift can take the well to depletion, the recoverable production may not be quite as much as using a more expensive beam pump system to pump liquids out of the well in the latter stages of depletion.
- Use of special standing valves with provisions to bleed off large liquid slugs may increase production.
- A new two-piece plunger concept is introduced that requires little or no shut-in period and also is able to operate with a packer present. Design parameters are being further developed.

REFERENCES

1. Lea, J. F., "Dynamic Analysis of Plunger Lift Operations," Tech. Paper SPE 10253 November, 1982, pp. 2617–2629.

2. Beeson, C. M., Knox, D. G., and Stoddard, J. H., "Part 1: The Plunger Lift Method of Oil Production," "Part 2: Constructing Nomographs to Simplify Calculations," "Part 3: How to User Nomographs to Estimate Performance," "Part 4: Examples Demonstrate Use of Nomographs," and "Part 5: Well Selection and Applications," Petroleum Engineer, 1957.

3. Otis Plunger Lift Technical Manual, 1991.

4. Lea, J. F., "Plunger Lift Versus Velocity Strings," *Journal of Energy Resources Technology,* Vol. 121, December 1999, pp. 234–240.

5. Ferguson Beauregard Plunger Operation Handbook, 1998.

6. Foss, D. L., and Gaul, R. B., "Plunger Lift Performance Criteria with Operating Experience-Ventura Field," *Drilling and Production Practice, API*, 1965, pp. 124–140.

7. Phillips, D., and Listiak, S., "How to Optimize Production From Plunger Lifted Systems, Pt. 2," *World Oil*, May 1991.

8. McCoy, J., Rowlan, L., and Podio, A. L., "Plunger lift Optimization by Monitoring and Analyzing Well High Frequency Acoustic Signals, Tubing Pressure and Casing Pressure," SPE 71083, presented at the SPE Rocky Mountain Petroleum Technology Conference in Keystone, CO, May, 2001, pp. 21–32.

9. Hacksma, J. D., "Users Guide to Predict plunger Lift Performance," Presented at Southwestern Petroleum Short Course, Lubbock, Texas, 1972.

10. White, G. W., "Combining the Technologies of Plunger Lift and Intermittent Gas Lift," Presented at the Annual American Institute Pacific Coast Joint Chapter Meeting Costa Mesa, California, October 22, 1981.

11. Rosina, L., "A Study of Plunger Lift Dynamics," Master Thesis, University of Tulsa, Petroleum Engineering, 1983.

12. Ferguson, P. L., and Beauregard, E., "How to Tell if Plunger Lift Will Work in Your Well," *World Oil*, August 1, 1985, pp. 33–37.

13. Wiggins, M., and Gasbarri, S., "A Dynamic Plunger Lift Model for Gas Wells," SPE 37422, Presented at the Oklahoma City Production Operations Symposium, 1997.

CHAPTER 8

USE OF FOAM TO DELIQUEFY GAS WELLS

8.1 INTRODUCTION

Foams have several applications in oil field operations. They are used as a circulation medium for drilling wells, well cleanouts, and as fracturing fluids. These applications differ slightly from the application of foam as a means of removing liquid from producing gas wells. The former applications involve generating the foam at the surface with controlled mixing and using only water. In gas well liquid removal applications, the liquid-gas-surfactant mixing must be accomplished downhole and often in the presence of both water and liquid hydrocarbons.

The principal benefit of foam as a gas well de-watering method is that liquid is held in the bubble film and exposed to more surface area, resulting in less gas slippage and a low-density mixture. The foam is effective in transporting the liquid to the surface in wells with very low gas rates when liquid holdup would otherwise result in sizable liquid accumulation and/or high multiphase flow pressure losses.

Figure 8-1 shows a laboratory test comparison of tubing pressure gradient with surfactants vs. flowing gradients when surfactants are not used. This figure generally illustrates the potential for pressure gradient reduction realized with foam and the approximate range of producing conditions where foam has application.

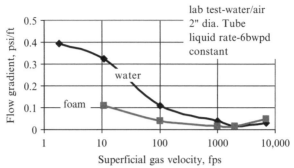

Figure 8-1. Producing gradients in lab column test at atmospheric pressure.

8.2 LIQUID REMOVAL PROCESS

Foam is a particular type of gas and liquid emulsion. Gas bubbles are separated from each other in foam by a liquid film. Surface active agents (surfactants) are generally used to reduce the surface tension of the liquid to enable more gas-liquid dispersion. The liquid film between bubbles has two surfactant layers back to back with liquid contained between them. This method of tying the liquid and gas together can be effective in removing liquid from low-volume gas wells.

Campbell et al.[1] describe the foam effect on production of liquids using the critical velocity. Equation 3-3 is repeated below:

$$V_t = \frac{1.593\sigma^{1/4}(\rho_l - \rho_g)^{1/4}}{\rho_g^{1/2}} ft/\sec \tag{8-1}$$

where σ = surface tension between the liquid and gas, dynes/ cm
ρ = density, lbm/ft^3

The subscripts l and g indicate liquid and gas. V_t is the terminal velocity or the gas critical velocity.

Campbell et al.[1] discuss that foam will reduce the surface tension and therefore reduce the required critical velocity. They indicate surface tension should be measured under dynamic conditions. They also discuss that foam will reduce the density of the liquid droplets to a complex structure containing formed water and/or condensate and gas. Thus, the beneficial effects of foam are described by the fact that the foamed liquid droplet density and surface tension both combine to reduce the required critical velocity.

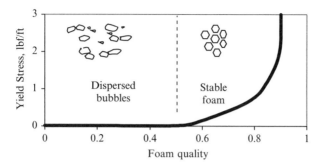

Figure 8-2. Yield stress characteristics of foam.

Water and liquid hydrocarbons react differently to surfactants. Liquid hydrocarbons do not foam well. This is particularly true for light condensate hydrocarbons. The gas-condensate bubble dispersion can be accomplished; however, the resulting foam is not stable and will readily separate. Light hydrocarbon liquids in general must be continuously agitated to maintain foaming.

One reason why hydrocarbons do not foam well is because hydrocarbon molecules are nonpolar and therefore have less molecular attraction forces between molecules. On the other hand, water molecules are polar and can build relatively high film strengths with surfactants. When both water and liquid hydrocarbons are present in the wellbore, foam is created mainly within the water phase, and the water foam helps carry along the liquid hydrocarbons. Laboratory observations indicate that when both water and light hydrocarbon liquids are present the liquid hydrocarbons tend to emulsify, and the foam is generated in the external water phase.

The percentage of gas in the foam mixture at operating pressure and temperature is termed foam quality (i.e., foam that is 80% gas is called 80 quality foam). Referring to Figure 8-2, the foam appears as a gas-liquid emulsion where the foam quality is less than approximately 50%. This type of foam or gas dispersion is unstable (i.e., gravity forces will separate the liquid and gas phases). At higher foam qualities, the liquid film becomes thinner and distorted because of surface tension. When the foam is nonflowing, it appears in the stable foam form as shown in Figure 8-2 and is relatively stable.

If this foam is caused to flow, a certain minimum stress will be required to overcome the interlocking of the bubble structures. This minimum stress is called a yield point. Thus, foams have an apparent viscosity, which is dependent on the shear rate operating in the moving stream.

8.2.1 Surface De-Foaming

If foam is used to unload a gas well, the foam must be broken before entering the separator and sales line. Breaking the foam at the surface is accomplished in several ways.

If not overly treated with surfactants, the foam will tend to break naturally if it is kept quiescent for a period of time. Liquid drains from the bubble film, and eventually the film ruptures and allows the gas to escape. This process is normal for foam dissipation.

This process may be aided by a further dilution of the surfactant concentration with produced or makeup water. Nonionic surfactants can be heated above their "cloud point" indicating a reduction in surfactant solubility, hence lowering the effective surfactant concentration. Surfactants can also be chemically treated with appropriate de-emulsifier chemicals of the opposite character. A completely stable foam may not be desirable because of the difficulty of de-forming in the separator. A surfactant that produces the maximum amount of foam that can be most easily processed may be the most desirable. Liquids are removed in the separator.

8.3 FOAM SELECTION

The application of foam to unloading low rate gas wells is generally governed by two operating limitations: economics and the success of foam surfactants in reducing bottomhole pressure. Both limits are defined by comparison to other methods of unloading wells.

Low-rate gas wells with producing GLRs between 1000 and 8000 scf/bbl are among the better candidates for foaming; although there is no real upper GLR limit. For high GLR wells, plunger lift may give better performance (i.e., produce with less bottomhole pressure than foam). Downhole pumps may be better suited for the lower GLR ranges and lower build-up pressure wells. Pumps require gas separation. The producing gradients expected with foam surfactants are ultimately controlled by the producing rates and well conditions and by the performance of specific surfactants in the well. Multiphase flow programs generally predict performance in gas wells where the foam is treated as the liquid (although it is both liquid and gas) in a two-phase system.

Laboratory tests tend to support the assumption that the liquid in the wellbore will form stable foam ranging between 50% and 85% quality under dynamic conditions. Foam quality appears to vary with the amount and type of liquids present as shown in Figure 8-3. The viscosity of foam varies with quality and with the amount and type of surfactant used. When quality is above approximately 52%, foam behaves as a plastic yield fluid having a plastic viscosity and yield strength (Figure 8-4). The yield strength is derived from the interlocking of gas bubbles and the strength of the bubble film. The viscosity of the foam is sensitive to the shear rate in the production string and could be

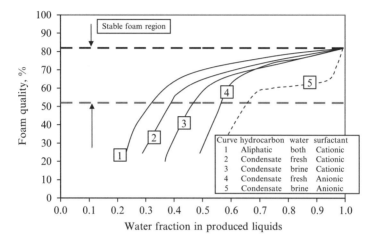

Figure 8-3. Foam quality in well systems.

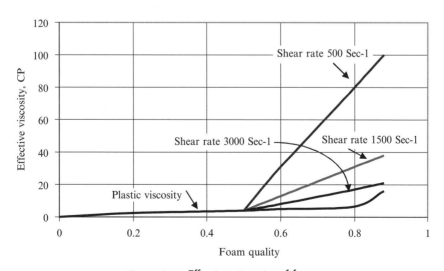

Figure 8-4. Effective viscosity of foam.

assumed to be predictable in the manner reported by Blauer et al.[2] and is as shown in Figure 8-4.

Other surfactant chemicals introduced into the wellbore can affect foam quality and viscosity. Corrosion inhibitors are packaged with surfactants to make them either water-soluble or water-dispersible, and they may also contain de-emulsifiers. Laboratory tests have evaluated the effect of two filming amine corrosion inhibitor compounds on foam quality. These tests show that the water dispersible inhibitor with de-emulsifier chemicals lowers foam quality (up to 10% lower), whereas the water-soluble inhibitor had little effect on foaming. The latter did tend to cause some oil and water emulsion problems.

The limiting economics for the application of foam are a function of both the chemical costs and the equipment cost. Chemical costs are proportional to the liquid (water) rate. At some level of water production, chemical costs will approach and exceed the cost of pumping. For example, a well being treated with surfactant costing $8.00/gal at a concentration of 0.1% will have a chemical cost of $34/bbl of treating fluids.

Foaming agents should not be used at locations where experience indicates that foam carryover and/or liquid emulsion treating problems are severe. Although the cost of surfactant injection equipment may be small, the cost for labor, chemicals, additional well equipment, or modification

Table 8-1

Advantages	Disadvantages
Foam is a very simple and inexpensive method for low-rate wells. Chemical costs are proportional to the liquid water rate.	Surfactant may result in foam carryover or liquid emulsion problems.
No downhole equipment is required. (However, a capillary injection system may be very beneficial to low-rate wells tending to produce in slugs.)	The foaming tendency for various systems depends on the amount and type of well fluids and on surfactant effectiveness. Well producing substantial condensate (i.e., more than 50% condensate) may not foam.
Method is applicable to wells with low gas rate where gas velocities may be on the order of 100 to 1000 fpm in the production string. The value is approximately 1000 fpm for critical velocity in unfoamed wells.	

of existing equipment to handle foam carryover and emulsions can be significant. Foam has also been used with plunger lift operations to improve the performance and reliability of plunger cycles. Surfactant use on plunger installations appears to be helpful when the gas rate is very low (plunger rise time is long) and/or when the gas rate and volume after plunger arrival is insufficient to clear the well of residual liquids. Surfactants are used to minimize gas slippage. This treatment applies to wells where the tubing-casing differential pressure remains relatively high after plunger arrival, thus indicating liquid under the plunger.

Table 8-1 lists some advantages and disadvantages for foam lifting of liquids that should be evaluated before selecting this method for unloading gas wells.

More information about selection testing is presented in the following material.

8.4 FOAM BASICS

This section provides background information on foam surfactants and the character of foams. The goal is to provide a better understanding of the liquid removal process and aid in the evaluation of information supplied by chemical companies.

8.4.1 Foam Generation

To produce useful foam, it is necessary to get a good dispersion of gas and liquid phases (foam generation) and then maintain the bubble film for a useful period (foam stability). Foam generation is accomplished through agitation of liquid with gas. This process is enhanced when the surface tension of the liquid is lowered so that gas may more easily disperse throughout the liquid phase. This is part of the role of surfactants. Water has a surface tension of approximately 72 dyne/cm, which is generally reduced to the 20 to 35 dyne/cm range with surfactants used for foaming. Liquid hydrocarbons also generally have surface tensions in the 20 to 30 dyne/cm range at lower pressures.

8.4.2 Foam Stability

Foams begin to deteriorate as soon as they are formed. Excess liquid between surfactant layers drains from the bubble film, resulting in thinning and weakening of the bubble wall. Liquids in the bubbles below are constantly replenished by the drainage from bubbles above. Also, the bubbles grow as the trapped gas expands until the liquid film becomes thin from drainage and expansion, and the film eventually breaks. Campbell et al.[1] describe the thinning of the foam film in terms of the critical micelle concentration (CMC). The CMC is the point at which the addition of surfactant molecules to a solution results in the formation of colloidal aggregates. When the foam formed above the structural disjoining, pressure is positive due to the presence of micelles in the film. The important factors for thinning rate are the surface rheology and film structure. The determining role is the film structure. Based on this model, the following effects are predicted. The more micelles present in solution, the easier the film ordering. The foamer with the lower CMC would at the same concentration have more micelles present and be more stable. If the foam is used at many times the CMC, the produced foam would be more stable.

Foam stability can be increased by reducing the liquid drainage rate and by increasing the resiliency of the surfactant layer.[3] Surface and bulk viscosity of the surfactant affects foam generation and stability. A higher viscosity will retard liquid drainage. However, high bulk viscosities are not attained in dilute solutions of most surfactants, and their surface viscosities are only moderate. Therefore, a "surface excess" (concentration of surfactant higher than the minimum required to start foaming) of the surfactant is of major importance in producing a stable foam. The surface excess also imparts a property termed "surface elasticity." As soon as the bubble expands, the excess concentration is decreased, and an increased surface tension that resists expansion will result; the reverse is true during contraction of the bubble. Thus the "surface excess" may be sufficient to maintain the surfactant film despite some local thinning of the surfactant concentration.

Surfactant effectiveness often passes through a maximum at intermediate surfactant concentrations.[3] A solution that is too dilute will not allow the range of surface effects (i.e., surface tension reduction, film elasticity, repair of ruptured bubbles, etc.) required for foaming. A solution that is too concentrated may cause excessive foam stiffness, high apparent foam viscosity, and/or excessive liquid-oil emulsions, as well as increasing the cost of treating the well. Laboratory tests indicate that many surfactants have an optimum effectiveness at approximately .1% to .2% concentration in the water phase. Campbell et al.[1] indicate that experience dictates a surfactant dosage of 1000–4000 ppm. Actually, the concentration should be based on the "active" quantity of the surfactant in the surfactant mixture as received from the supplier. A surfactant that is 50% active will have .5 lbs of active ingredient per pound of surfactant. The optimum effectiveness, therefore, may be at .05% to .1% "active" concentration in the well fluids. In water–hydrocarbon liquid mixtures, the optimum water treatment rate can be applied to the total liquid rate. This allows for some surfactant loss to emulsion droplets.

8.4.3 Surfactant Types

Surfactant molecules have a water-soluble (hydrophylic) end and a non–water-soluble (hydrophobic) end. Thus, a surfactant contains hydrophylic and lipophylic (oil soluble) constituents, which cause the molecule to concentrate at the interface between the water and nonwater phases. When the concentration of the surfactant is such that the interface surface area is completely covered with a maximum number of surfactant molecules, the solute is said to be at its critical concentration. Subsequent additions of the surfactant must enter one of the liquid phases. Therefore, the solubility of the surfactant must be capable of providing a surfactant concentration that will supply surfactants for the large surface area created by dispersed bubbles. Additionally, there are ionic attraction (polar) forces acting between surfactant and liquid molecules that strengthen the film.

Surfactants can be classified according to their ionic characters as nonionic, cationic, or anionic.[2] A description of these classifications follows.

Nonionic surfactants

Nonionic surfactants are typically polyoxyethylated compounds of phenols or alcohols. Although the solubility of most detergents in water increases with temperature, nonionic types are generally more soluble in cool water. When heated, the surfactant loses solubility, and the solution becomes cloudy (cloud point). High concentrations of salt and high temperatures decrease the solubility of polyoxyethylated detergents, thus lowering their cloud points.

Therefore, members of this series of surfactants with the higher ethyloxy contents (more water soluble) should be used in saline waters. Polyoxyethylated surfactants are available in homologous series ranging from oil-soluble to water-soluble types. Because they are nonionic, they are relatively unaffected by the activity or chemical nature of the formation brines, and they are used in wells of unknown brine character. They tend to cause less emulsion problems than do ionic surfactants. Heating the produced foam above the cloud point (approximately 150°F) helps break the foam.

Anionic surfactants

Anionic surfactants are excellent water foamers. Anionic surfactants are typically nonionic type foamers that have undergone a sulfation process during manufacturing. The addition of the sulfate radical (SO_4) onto the molecule causes the surfactant to become more polar and anionic in character and increases its solubility in water.

As with nonionic surfactants, they are available in a homologous series in which the mid-range oil–water solubility cut (10 to 12 carbon atoms) is generally preferred. Some anionic surfactants may be adversely affected by high brine solutions. See Libson and Henry[4] for a successful case history use of a particular chemical classified as an anionic surfactant.

Cationic surfactants

Cationic surfactants, such as quaternary ammonium compounds, are effective foaming agents, often performing more effectively in brines than in freshwater. The low-molecular-weight agents of this type are among the most effective agents for foaming mixtures of hydrocarbons and brines and reportedly high-molecular-weight quaternaries show some effectiveness in foaming wellbore fluids with high percentages of liquid hydrocarbons. On the other hand, high-molecular-weight quaternaries show a decrease in effectiveness in brine solutions. In some cases, particularly with overtreating, oil and water emulsion problems are created. Cationic surfactant performance data are shown in Figures 8-5 and 8-6.

Amphoteric surfactants were found to be good foam surfactants in one laboratory study.[1] Amphoteric compounds exhibit cationic character in an acidic solution, anionic character in basic solutions, and nonionic character in neutral solutions. They are reportedly very good foaming agents in high (200°F) temperature test with up to 10% salt in solution. At 70°F, their performance in laboratory foam column tests was slightly less favorable than anionic or cationic surfactants. Some amphoteric surfactant performance data are shown in Figures 8-5 and 8-6 and are representative of this type of surfactant at room temperatures.

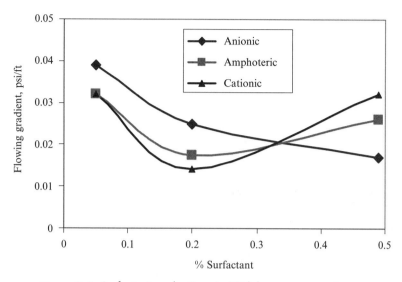

Figure 8-5. Surfactant evaluations in 27 laboratory test column.

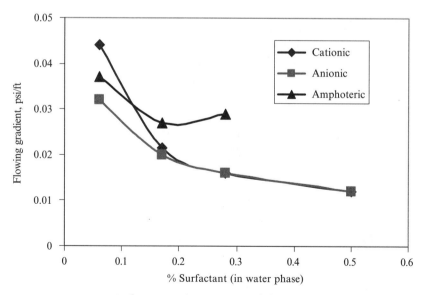

Figure 8-6. Surfactant evaluations in 27 laboratory test column.

Foaming agents for hydrocarbons

In general, hydrocarbons are difficult to foam especially without the presence of water. Liquids with less than 70% to 80% hydrocarbons with water may be foamed more easily than liquids with higher percentages of hydrocarbons. For higher percentages of hydrocarbons, more expensive foaming agents such as fluorocarbon surfactants have been used in the past. Experience shows that most liquids from gas wells are either water or water with a lower percentage of hydrocarbons.

8.4.4 Foaming with Brine/Condensate Mixtures

Some water–hydrocarbon liquid mixtures show the ability to foam, whereas other mixtures do not. Generally speaking, only the water phase in a water-hydrocarbon mixture produces stable foam (i.e., foam where the bubble film is sufficiently strong to hold the water and gas in a bubble structure at significantly high gas concentrations). This is because the water molecule is polar and permits a polar hydrogen bond with these surfactants. Foam surfactant molecules have an oil-loving and a water-loving end, which becomes oriented at the interface and strengthens the interface film.

Hydrocarbons do not foam well because there are no polar bonds with the surfactants; indeed, the principal mechanism for hydrocarbon foam surfactant is to place high molecular polymers at the interface to interact with many oil molecules to build viscosity and/or molecular attraction between several oil molecules. Although this provides some development of film strength, it is significantly weaker than water foam films. To successfully foam a well, it is necessary to obtain an effective foam condition in the water phase. If free-oil is present, it may be "lifted" by the drag forces of the water foam rather than by a foaming tendency in the oil itself.

Effect of condensate (Aromatic) fraction

A certain amount of the surfactant will react in the system to create an oil-in-water emulsion. Laboratory data show that the emulsion tendency is much higher for low-molecular-weight aromatic and cyclic hydrocarbons than for aliphatic hydrocarbons. As would be expected, an increase in the aromatic (principally toluene) content diminishes the foaming tendency in oil-water mixtures.

Figures 8-7 and 8-8 show results of laboratory tests for the column pressure gradient and estimated foam qualities for various mixtures of

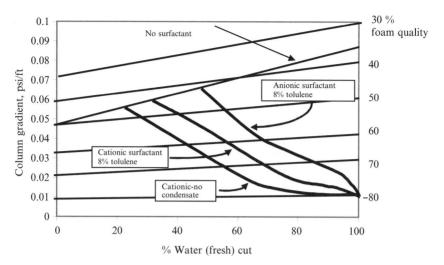

Figure 8-7. Foam column test with condensate and fresh water system superficial gas velocity at 60 fpm.

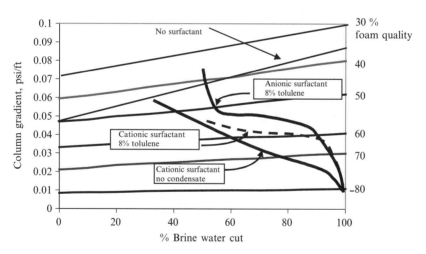

Figure 8-8. Foam column test with condensate and brine with superficial gas velocity at 60 fpm.

hydrocarbon liquids in both fresh and brine water. In these figures the foam quality and column gradient are for the same conditions. The foam quality correlation is based on the gradient observed and on some measurements of foam quality obtained by shutting in the test column and measuring the volume of foam and subsequently the liquid components of the dissipated foam.

   Although the tests were run at very low superficial gas velocities (which may not characterize the true gradient in a field well), the low velocity does diminish the effect of flowing friction and allows for a better correlation with foam quality. Based on the data shown in these figures, the foaming tendency of a mixture is good, even to relatively low water cuts, for mixtures containing hydrocarbon liquids with no toluene content. However, one of the hydrocarbon liquids used to represent aromatic content (Stoddard solvent) contained 16% heavier aromatics but no benzene or toluene. On the other hand, mixtures with varied light end aromatic content showed some increase in column gradient as the toluene content increased.

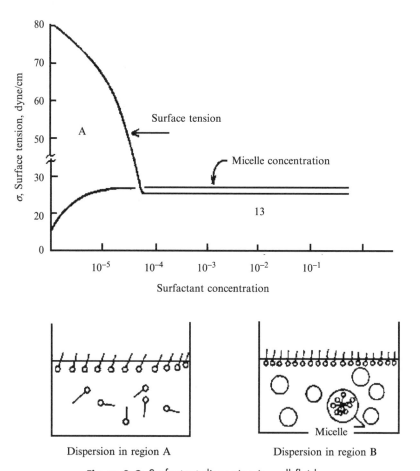

Dispersion in region A   Dispersion in region B

Figure 8-9. Surfactant dispersion in well fluids.

Figure 8-9 shows a pictorial dispersion pattern for foam surfactants in oil-water systems. As a surfactant is added, the surface tension of the water decreases to the point where the entire surface is saturated by a surfactant. Beyond this point, the surface tension is relatively constant at some low value characterized by surfactant constituents. Also, as surfactant concentration increases, more surfactant molecules enter the water phase and, at a critical concentration, the heretofore dispersed surfactant molecules begin to conglomerate in clusters or micelles. This point is reached at a much lower surfactant concentration than that required to stabilize the surface tension and effective foaming. The net result of these interactions in liquid is that a greater abundance of oil is contained in the bubble film, which reduces the film strength and, consequently, foam quality. Also, apparent viscosity of the liquid can increase because of emulsions; this is particularly evident in the mid-range water cut mixtures that exhibited higher pressure gradients than other tests where the liquid contained no surfactant or aromatic hydrocarbons.

Figure 8-10 is a general correlation[5] of percentage aromatics contained inhydrocarbon liquids as a function of specific gravity and average molecular weight of the C7+ constituents. This can be used when an

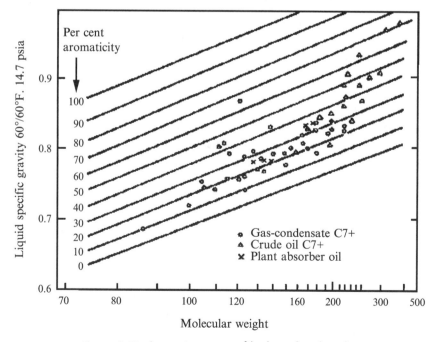

Figure 8-10. Aromatic content of hydrocarbon liquids.

analysis of the condensate for aromatic content has not been run. Three condensates were used in the laboratory tests. All showed approximately 25% total aromatic content, including 8% to 12% benzene plus toluene fractions.

Effect of brine

Brine-oil mixtures are associated with field locations where the ability to generate foam has been questioned. However, tests indicate that brine with no oil phase present will foam about as well as fresh water. Effective foam quality in an oil-water mixture decreases more rapidly, however, when water salinity is high. In comparing Figures 8-7 and 8-8, the effect of brine rather than fresh water is seen to be a reduced foaming tendency. This effect was more drastic with the anionic surfactant tested compared with the cationic surfactant. Two things account for brine reaction behavior:

- Salts tend to reduce the solubility of surfactants in water.
- The critical micelle concentration (concentration of surfactant when micelle formation begins) is reduced.

The water-attracting ends of the surfactant molecule see a reduced number of unassociated water molecules (those without ionic attraction to the disassociated salt ions) to attract the surfactant molecule. As the concentration of surfactant increases, the surfactant molecules form micelles (clusters) in the water; each with the oil-attracting ends attracted to the center of the cluster. Thus, in brine waters, there tends to be a larger number of surfactant micelles dispersed throughout the fluid. Some of the free hydrocarbon liquid is drawn into the micelle cluster, which causes a greater quantity of oil to be in the water phase.

Investigations by Hough et al.[6] indicate that the surfactant tension in the methane-water system is slightly lower than the air-water system. Surface tension also becomes smaller as the pressure of the system increases because of the gas driven into the liquid. Also, a slight increase is noted in the surface tension of water as salinity increases.[5]

Brine solutions alter the emulsion tendency for some surfactant-hydrocarbon liquids. The surface tension for wellbore produced fluids at relatively low pressures should not be significantly different from the air and liquid systems investigated, and the critical control is in the surfactant treatment.

8.5 OPERATING CONSIDERATIONS

8.5.1 Surfactant Selection

Many surfactants have undergone screening tests in the laboratory and at field locations.[1-4] The results of these tests indicate which cationic, anionic, and amphoteric surfactants show the best performance. However, formation water and liquid hydrocarbons differ between fields, and/or other surfactants may also be effective.

Before using surfactants in the field, a field screening test should be performed using samples of produced liquids. Field screening tests show how well the agent foams the produced liquids and the compatibility of the surfactant and produced fluids. The effectiveness of de-emulsifier chemicals can also be checked.

8.5.2 Bureau of Mines Testing Procedures

Some simple testing procedures used to determine what surfactant will work best in your well fluids are outlined by Dunning et al.[3]

Figure 8-11 shows a testing procedure where the largest amount of foam carried over to a collection beaker indicates the best surfactants. The test is conducted by placing a small sample of wellbore liquid into a tube with a specified amount of foaming agent. Gas is then injected into the bottom of the tube through a fretted disk at a specified flow rate. Liquid carried by the foam through the tube is collected and weighed. Foaming agents are thus compared to determine which most effectively carries the liquids from the tube over time. The Bureau of Mines test is simple, quick, and has a relatively low cost. It allows for a quick evaluation of several candidate foaming agents before expensive field trials are commenced.

Figure 8-12 illustrates a test where the foaming agent is allowed to drop 90 cm, and foam heights are measured with time as the test proceeds.

Vosika[8] describes the successful use of the above two test procedures and successful application of the screening tests in the field. As described above, Campbell et al.[1] discuss how surface tension and density of foamed liquid droplets have the beneficial effect of lowering the required critical velocity of gas. As such, they focus on laboratory testing to determine the dynamic surface tension and the density of the foam. They use the maximum bubble pressure method to determine surface tension. A small glass capillary (0.25 mm diameter) is immersed into the fluid of

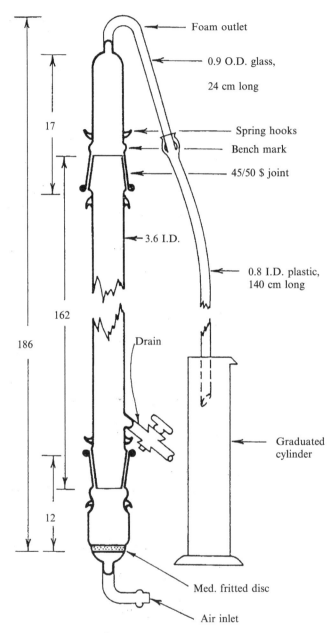

Figure 8-11. Bureau of Mines[3] setup for dynamic testing of foaming agents. (Dimensions are in centimeters.)

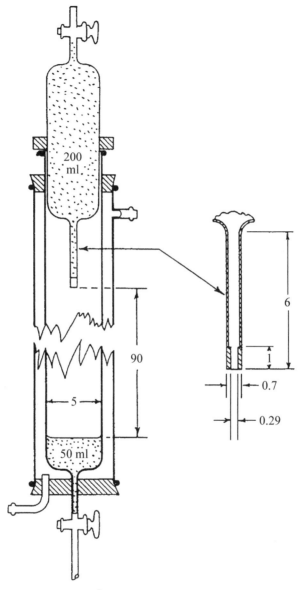

Figure 8-12. Bureau of Mines[3] setup (Ross-Miles Apparatus) for static testing of foaming agents. (Dimensions are in centimeters.)

interest at a controlled temperature. To simulate production, the brines are kept under a carbon dioxide sparge to maintain the pH and prevent solid precipitation. Nitrogen is bubbled into the solution at a fixed flow rate and the pressure for bubble detachment is measured. To correct for differences in immersion depth, a larger glass capillary (4.0 mm diameter) is also immersed in the solution and the detachment bubble pressure is used as a reference. To examine the dynamic effect, the flow rate of the nitrogen is varied using a mass flow controller with a range between 1 to 100 bbl/sec. The surface tension is strongly a function of the dynamic bubble rate in the concentration of interest (>1000 ppm). Other details may be found in Reference 1.

Continuing the discussion of the testing described by Campbell et al.,[1] stability is tested using a blender test. A volume of 100 mL of fluid either synthetic or produced fluid at the condensate/water ratio of the well's fluids is agitated at a low speed for 60 seconds. The volume of total fluids and foam is immediately recorded. The time for 50 mL of water to separate from the foam is also recorded. The test is run at room temperature. The foam density is recorded as a function of concentration of various surfactants.

8.5.3 Unloading Techniques and Equipment

Wells are unloaded with surfactants using two techniques: batch (single event) treatment or continuous surfactant injection.

Batch treatment

The surfactant quantity is based on an estimate of the liquid to be unloaded. The well is usually shut-in, and the liquid load can be determined from the casing-tubing differential. The appropriate surfactant quantity for a 1% surfactant concentration is mixed in 20 gallons or more of produced fluid or water and is then pumped or lubricated into the tubing. Foam "soap sticks" can be used instead of liquid surfactants. The well is then opened to flow. Batch treatments are best suited to wells that require liquid unloading on an infrequent basis because of the time required to perform batch treating.

Continuous foam injection

Using an appropriate equipment arrangement (Figure 8-13), a quantity of surfactant is continuously injected into the wellbore where it mixes with the produced liquid and gas to generate foam. Surfactant injection can be either down the casing-tubing annulus or through the tubing; the production travels through the alternate path.

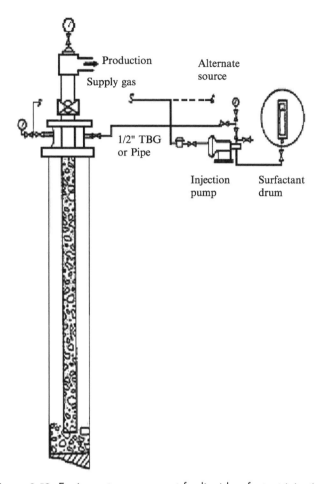

Figure 8-13. Equipment arrangement for liquid surfactant injection.

The equipment required for a foaming operation is shown in Figure 8-13. Most locations use a gas-powered pump operating off one side of the casing. A pump can be located under a chemical barrel so that the entire pump system can be covered or insulated.

Because the surfactant may tend to hang on the tubular walls, the surfactant solution should be diluted to provide a larger volume to ensure reaching the fluid level, or it should be washed down with additional fluid after surfactant injection. The surfactant should be diluted to approximately 1 part in 10 parts of water so that a surfactant dosage of 2 quarts of diluted solution per barrel of produced fluid is injected for a surfactant concentration of approximately 0.1%.

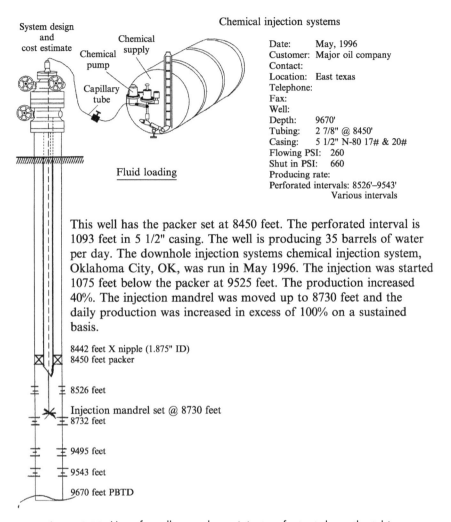

System design
and
cost estimate

Chemical
pump

Chemical
supply

Capillary
tube

Fluid loading

Chemical injection systems

Date: May, 1996
Customer: Major oil company
Contact:
Location: East texas
Telephone:
Fax:
Well:
Depth: 9670'
Tubing: 2 7/8" @ 8450'
Casing: 5 1/2" N-80 17# & 20#
Flowing PSI: 260
Shut in PSI: 660
Producing rate:
Perforated intervals: 8526'–9543'
                      Various intervals

This well has the packer set at 8450 feet. The perforated interval is 1093 feet in 5 1/2" casing. The well is producing 35 barrels of water per day. The downhole injection systems chemical injection system, Oklahoma City, OK, was run in May 1996. The injection was started 1075 feet below the packer at 9525 feet. The production increased 40%. The injection mandrel was moved up to 8730 feet and the daily production was increased in excess of 100% on a sustained basis.

8442 feet X nipple (1.875" ID)
8450 feet packer

8526 feet

Injection mandrel set @ 8730 feet
8732 feet

9495 feet

9543 feet

9670 feet PBTD

Figure 8-14. Use of small cap tube to inject surfactant down the tubing.

In winter operations, the diluting liquid can be 50% ethylene glycol. A better but more expensive injection system is one using a capillary tube strapped to the outside of the production tubing. A Sperry-Sun chemical injector system is but one example of this type of equipment arrangement. The advantage of this equipment is that relatively small quantities of surfactant can be injected with assurance that it will reach the desired downhole injection point. This circumvents problems with fluctuating fluid levels in the annulus. Figure 8-14 shows a similar system where the surfactant is injected down a capillary string lubricated inside the tubing.

The tubing should be landed approximately in the top 1/3 of the completion interval. Where the tubing is landed can be controversial for large pay section thicknesses. However, landing below the top of the pay should increase draw-down. Although production flow is generally through the tubing, it can be through either the tubing or casing annulus; however, possible casing corrosion must be considered.

A selection of which path to use is based on which conduit will give gas velocities in the range of 3 to 12 fps at operating pressure and temperature. There may be some liquid (foam) holdup in larger cross-sectional flow areas (where velocity is in the lower range), particularly when the foam quality is low because of poor surfactant performance or the presence of significant liquid hydrocarbons. In this case, it is probably better to flow up the tubing.

A multiphase flow pressure loss computer program based on the foam system is recommended for estimating producing gradients and for selecting the optimum flow path.

To initiate injection, it is advantageous to perform a batch treatment[8] with a high foamer concentration before continuous injection. This will unload existing wellbore liquids so that lower injected concentrations will be more effective. By dumping the concentrated foamer (the amount to be determined by the estimated liquid in the wellbore and the effective concentration of the foamer being used) down the annulus and "stop-cocking" or sometimes called "stop-clocking" (intermittently producing and then shutting in) the well, good mixing should occur with the accumulated liquids. Foamer injection should begin after the initial unloading of the liquids.

8.5.4 Determining Surface Surfactant Concentration

To determine the correct concentration of the foamer in the surface tank, the following needs to be considered[8]:

1. Minimum effective foam concentration (generally between .1 and .5% for foamers)
2. Estimated amount of liquid produced by the well
3. Injection rate

The surface concentration can be determined (as an example) as follows:

$$C_S = C_E \frac{L_P + L_I}{L_I}$$

where C_S = surface concentration
C_E = minimum effective concentration
L_P = liquid produced
L_I = liquid injected

Example 8.1

Minimum effective concentration = .2%
Amount of produced liquid = 4 bbl/day water + 1 bbl/day condensate
= 5 bbl/day = 210 gal/day
Injection rate = 20 gal/day
Then the required surface concentration is

$$C_S = 0.2 \frac{210 + 20}{20} = 2.3\%$$

The minimum effective concentration used in this calculation should be multiplied by a factor of two or three to determine the initial concentration at the surface used to compensate for uncertainties. Again, this is but an example technique used in one location.[8]

After injection begins and production increases and stabilizes, it is then time to optimize the amount of foamer used. Concentrations should be decreased until a maximum production rate is obtained using a minimum amount of foamer. This is just one example of how this might be done.

If a packer is present in the well, the method illustrated in Figure 8-14 provides a way to lubricate a small diameter capillary tube down the tubing to inject chemicals into the well without removing the packer. Figure 8-15 shows this system being installed. Two suppliers of this service are Downhole Injection Systems (Oklahoma City, OK) or Dyna-Coil division of Dyna-Test (Kilgore, TX). Figure 8-16 shows results from one well where this system was installed. In this case, this system with the surfactant showed obvious improvement.

Surfactants can also be added to the well in the form of soap sticks. The apparatus in Figure 8-17 provides an automated method of dropping soap sticks into a wellbore as developed by one company.[9] This method has recently been enhanced with a computerized system to monitor well performance and drop the sticks as required to optimize production.

Pro-Seal Lift offers a device that has the sticks arranged in a rotating cylinder so that the sticks are side by side and do not have to have one on top of the other.

Figure 8-15. Capillary tube system to inject surfactants being lubricated into well.[10]

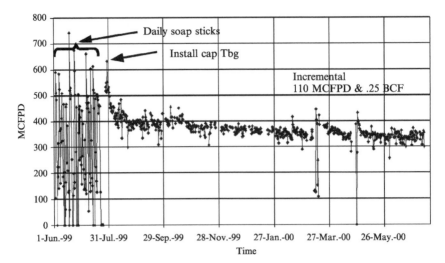

Figure 8-16. Case history in 2 3/8-inch tubing with packer injecting surfactant with capillary tubing system to bottom of the tubing.[10]

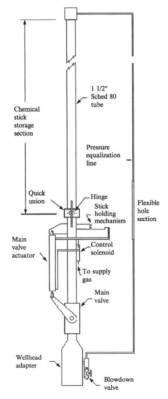

Figure 8-17. Automatic soap stick dispenser apparatus.[9]

8.5.5 Instrumentation

To evaluate the performance of a well using foam surfactants, it is necessary to have casing and tubing pressures and production data. In a continuously flowing well, the fluid level in the annulus will be at the bottom of the tubing (assuming no packer and no gas path from the top of the casing). The casing-tubing differential pressure then reflects the pressure gradient in the tubing neglecting some correction for the weight of the gas in the annulus.

The producing bottomhole pressure can be found by accounting for the gas gradient in the casing and the surface casing pressure. The fluid level will always be at the tubing intake depth in the flowing well, and the casing is shut-in because downhole gas separation will fill the annulus with gas. However, if the producing rate is so low that slugging or annulus cycling occurs, the pressure equilibrium established between the annulus and the tubing may be temporarily upset in proportion to the size of the pressure surges. As a result, some liquid or foam accumulation may occur above the tubing intake point when the annulus pressure is recovering from a de-pressuring cycle after the production of a liquid slug.

8.5.6 Chemical Treatment Problems

Emulsion problems

Some success in breaking the emulsions has been realized with de-emulsifier chemicals. Chemical companies usually recommend appropriate de-emulsifier chemicals, and they should be evaluated in bottle tests before using in the well. The chemical is injected to the line upstream of the separator to allow mixing with produced liquids before entering the vessel.

In some cases, wells are treated with other surface-active compounds, such as those packaged with corrosion inhibitors. The surfactants may contribute to emulsion or foam stability problems, and either the foam surfactant or the other treatment chemical will have to be changed or the treatment dosage will have to be moderated to improve the situation. Further, it may be necessary to discontinue inhibitor treatment for a while to be more definite about the cause of severe emulsion problems.

Foam carryover

Foam carryover into lines and separators sometimes causes upsets and interferes with level controls. De-foamer chemicals can be effective in suppressing the foam. The de-foamer is injected into the flowline upstream of the separator to allow for mixing before the stream reaches the separator. Selecting the type and treatment rate depends on several local factors, and tests should be conducted on location with a chemical company representative.

When the foam is broken, the liquid phases are separated in the production separator where the produced fluid is in as near a quiescent state as possible. Therefore, the separator should be relatively large, hopefully allowing gas velocities approximately 2 fps and 5 minutes or more holding time for liquids.

Liquid taken from the separator may still have significant oil-water emulsions. If this occurs, it will be desirable to allow additional separation time in a liquid holding tank. If the separator is a three-phase separator, it may be desirable to discharge the water to a holding tank if significant emulsions would otherwise remain in the water going to the pit.

If foam carryover or persistent emulsions continue with existing separation equipment, it may be desirable to chemically treat the produced stream to destroy the foam surfactant activity.

8.6 SUMMARY

- Foaming is usually possible for liquids in a gas well production stream if the liquids are water or a high percent of water.
- Foaming is more difficult and expensive for hydrocarbon with water percentages less than about 80%. Above that, foaming is more easily accomplished. Most gas wells produce either water or water and lower percents of hydrocarbons.
- Different surfactants should be tested on a sample of wellbore fluid to determine the most effective foamer using the Bureau of Mines[3] testing procedures or other testing procedures[1] outlined here.
- Foaming may assist other lift methods such as plunger lift. Foaming may also be used to stabilize a well that is producing erratically.

If no packer is present, surfactants can be introduced down the annulus either with continuous or batch treatments, usually mixed with water.

Another method is automatic or manual launching of solid soap sticks periodically down the tubing. If a packer is present, a capillary tube system of injecting at or below the end of the tubing is available and has been successful in many cases.

REFERENCES

1. Campbell, S., Ramachandran, S., and Bartrip, K., "Corrosion Inhibition/Foamer Combination Treatment to Enhance Gas Production," SPE Paper 67325, presented at the SPE Production and Operations Symposium, Oklahoma City, OK, March 24–27, 2001.
2. Blauer, R. E., Mitchell, B. J., and Kohlhaas, C. A., "Determination of Laminar, Turbulent, and Transitional Foam Flow Losses in Pipes," SPE Paper 4888. Presented at the 44th Annual California Regional Meeting, April 4–5, 1974.
3. Dunning, H. N., Eakin, J. L., and Walker, C. J., "Using Foaming Agents for Removal of Liquids from Gas Wells," Monograph 11, Bureau of Mines, Am. Gas Assoc., New York, NY, 1961.
4. Libson, T. N., and Henry, J. R., "Case Histories: Identification of and Remedial Action for Liquid Loading in Gas Wells...Intermediate Shelf Gas Play," SPE Paper 7467, presented at 53rd Annual Fall Meeting of SPE of AIME, Houston, TX, October 1–3, 1978.
5. Jacoby, R. H. NGPA Phase Equilibrium Project, API Proceedings Division of Refining, 1964, p. 288.
6. Hough, E. W. et al., "Interfacial Tensions at Reservoir Pressures and Temperatures; Apparatus and the Water-Methane System," Trans. AIME, 1951, p. 57.
7. Katz, D. L. et al., Handbook of Natural Gas Engineering, McGraw Hill, New York, 1959.
8. Vosika, J. L., "Use of Foaming Agents to Alleviate Liquid Loading in Greater Green River TFG Wells," SPE/DOE 11644, presented at the 1983 SPE/DOE Symposium on Low Permeability, Denver, CO, March 14–16, 1983.
9. Sloan, R., High Plains Wireline, Elk City, OK., Waggoner, Richard, Foamtech, Woodward OK, 1966.
10. Letz, R. S., "Capillary Strings to Inject Surfactants," SWPSC School on De-Watering Gas Wells, Lubbock, TX, April 24, 2001.

HYDRAULIC PUMPS

9.1 INTRODUCTION

Hydraulic-powered downhole pumps are powered by a stream of high-pressure power fluid (water or oil) supplied by a power fluid (PF) pump at the surface and sent to a downhole pump or pump engine.
Hydraulic pumps are basically of two types:

1. Piston downhole pumps that are similar to beam down-hole pumps
2. Jet downhole pumps that reduce the pressure on the formation by high-speed power fluid flow through the throat of a venturi or jet pump nozzle-diffuser combination

The surface PF pump can be either a piston type or a centrifugal high-pressure pump with sufficient capacity to transmit a pressurized power fluid to the downhole pump. Power is supplied to the PF surface pump by either an electric- or gas-powered prime mover.
Figure 9-1 shows an approximate depth-rate application chart for jet and reciprocating hydraulic systems application.
Figure 9-2 shows a typical skid-mounted, electric-powered triplex PF pump. Ordinarily the PF pump is either a split-case centrifugal pump, a horizontal electric pump (ESP) unit, or a triplex piston pump. Centrifugal PF pumps tolerate solids better but are less efficient than triplex positive displacement pumps.
The power fluid transfers the power necessary to lift liquids from the surface to the bottomhole pump. The power fluid can be oil or water. Generally, two types of power fluid systems are used with hydraulic pumps: One where the power fluid is kept isolated from the production

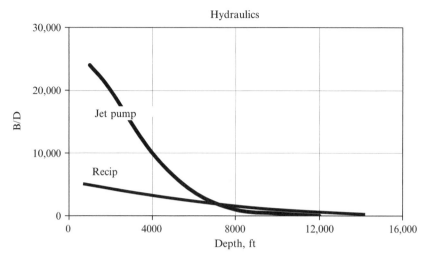

Figure 9-1. Approximate depth-rate envelope for jet and reciprocating hydraulic pumping systems.

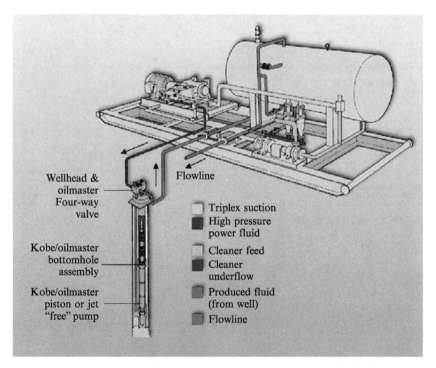

Figure 9-2. Hydraulic pumping system. (Courtesy Weatherford.)

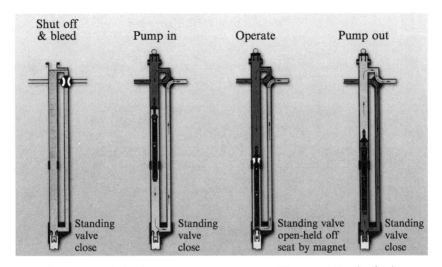

Figure 9-3. Hydraulic pumps—open systems. (Courtesy Weatherford.)

fluids and one where the two are allowed to mix. The latter system (where the power fluid is combined with the production fluids) is called an "open system" and is more common. There is considerable literature[1-5] discussing the features of hydraulic pumping and how it may be applied to dewatering gas wells. Figure 9-3 shows operations with a hydraulic pump using the four-way valve at the surface. With the valve in the four different positions, the following four functions can be performed:

1. In the first illustration, the system is bled down to low pressures.
2. The pump is then "pumped-in" by applying power fluid pressure over the pump.
3. Next the system is operated by injecting power fluid and pumping well and power fluid to the surface up the production tubing.
4. Finally, power fluid is reversed down the production line, and the pump is brought to the surface for inspection and/or replacement.

Figure 9-4 shows the surface facilities for the open power fluid system showing combined fluid coming from the wellhead.

Figure 9-5 shows a closed power fluid system with separate lines for production and power fluid.

The bottomhole pump can be either a piston pump, which is actuated by a pressured piston engine on top of the pump, or a jet pump, where the power fluid goes through a nozzle jetting into a throat and then

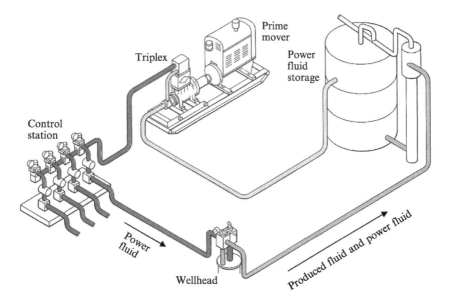

Figure 9-4. Surface facilities for an open power fluid system. (Courtesy Weatherford.)

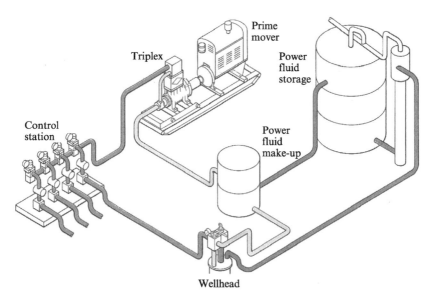

Figure 9-5. Closed power fluid surface facilities. (Courtesy Weatherford.)

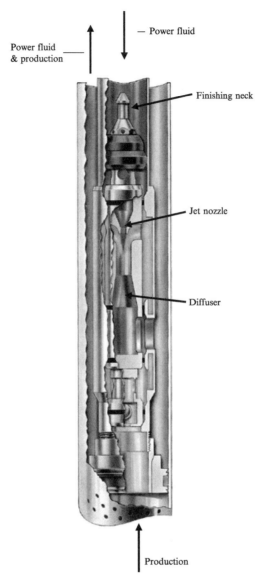

Power fluid

— Power fluid

Power fluid & production

Finishing neck

Jet nozzle

Diffuser

Production

Figure 9-6. Jet pump: One tubing string application.

diffuser creating a low-pressure area exposed to the formation. Figure 9-6 shows a typical jet pump with power fluid being supplied down the tubing and production and power fluid returning to the surface through the casing annulus. This particular configuration is of a free pump, designed to be pumpable to the surface by reverse circulation. This pump

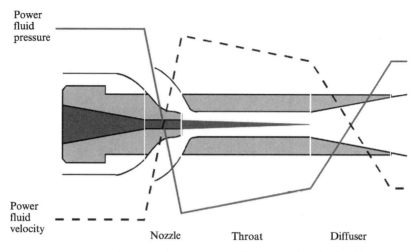

Power
fluid
pressure

Power
fluid
velocity

Nozzle   Throat   Diffuser

Figure 9-7. Jet pump throat section. (Courtesy Weatherford.)

is also equipped with a fishing neck for wireline retrieval in the event that the pump is stuck and cannot be reverse-circulated to the surface.

The jet pump is generally less efficient than the piston pump. The efficiency of the jet pump typically ranges from 15% to 25%, whereas a piston pump can be as high as 75%. The efficiency mentioned here is a ratio of power actually supplied to lifting fluids divided by total output power at the surface. The jet pump is much simpler, however, in design and is capable of tolerating solids better than the piston pump. The jet pump operates by the power fluid, creating a low-pressure area in the throat of the jet pump, which draws in the production from the well and carries it out a diffuser section, along with the power fluid. A schematic of the throat section of the jet pump is shown in Figure 9-7. Various combinations of throat sizes and nozzle sizes (having different offsets of the nozzle from the throat) are used to obtain a variety of production rates given various levels of power fluid pressure. In general, a big nozzle diameter flowing into a small throat provides the highest lift, whereas a small nozzle flowing into a larger throat allows for more volume. A jet pump always makes production and power fluid, so it always is an "open" system.

9.2 ADVANTAGES AND DISADVANTAGES

An illustration of a typical reciprocating hydraulic pump is shown in Figure 9-8.

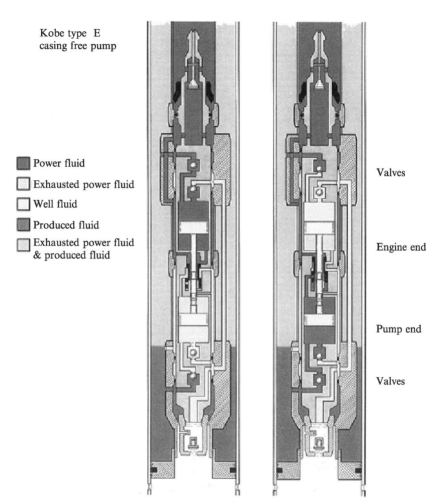

Kobe type E
casing free pump

Power fluid

Exhausted power fluid

Well fluid

Produced fluid

Exhausted power fluid
& produced fluid

Valves

Engine end

Pump end

Valves

Figure 9-8. Internals of a type E reciprocating hydraulic pump showing the engine end and the pump end.

A reciprocating hydraulic pump has ball and seat valves somewhat similar to a downhole beam pump. The pump end of the pump is driven by the engine end of the downhole pump assembly. The downhole stroke length is shorter (approximately 10 to 50 inches) than for a beam system and the SPM is much higher (approximately 20 to >100); however, these numbers for stroke length and operating SPM vary considerably with equipment models and sizes. The reciprocating pump can wear quickly when solids are present, and gas in the pump quickly reduces the efficiency of the system.

Table 9-1
Hydraulic Pump Advantages and Disadvantages

Advantages	Disadvantages
Piston pump can pump a well down to low pressures. However, the jet pump does not pull the well down to low pressures.	Requires a second string to vent gas or for power fluid. Closed vented system may have three lines downhole.
Easily adjustable at surface to change in well productivity.	Rate limited in 7-inch casing at approximately 1000 bbl/day (as long as gas has to be vented).
Can be used with closely spaced wellheads.	Power oil systems are an environmental and safety hazard. Will switch over to water after breakthrough.
Crooked holes are minimal problem.	Power fluid treating required to extend the life of surface and downhole pumps.
Good power efficiency with a reciprocating pump ($\approx 50\%$ +/−). Lower efficiency ($\approx 20\%$ +/−) with jet system.	Difficult to obtain good well tests in low-volume wells.
Power source can be remotely located.	Cannot run production logs while pumping.
Flexible; can usually match well production as well declines.	Vented installations are more expensive because more tubulars are required.
Paraffin easily prevented by heating or chemically treating the power fluid.	Treating for scale below packer is difficult.
Downhole pumps can be circulated out or retrieved by wireline when worn.	High operating cost if Triplex pumps are used to pressurize the power fluid for individual wells.
Adjustable gear box offers more flexibility for Triplex power fluid pump systems.	Gassy wells usually have lower volumetric efficiency and shorter pump life.
	Wellhead will freeze during shut-in for high WOR wells.

Although the reciprocating pump would be considered more delicate and more prone to failure compared with the jet pump, the reciprocating pump could bring the well pumping formation pressure to a lower level compared with a jet pump if no serious operational problems are present or if the run-life is not too short.

Table 9-1 lists some advantages and disadvantages of jet or piston hydraulic pumps. As indicated, hydraulic pumps have a particular range of application. Although the comments apply to both jet and piston type pumps, the jet pump cannot lower the flowing bottomhole pressure in the well as low as a piston pump. Because of gas interference, significant gas production will decrease the pump efficiency compared with pumping gas-free liquids.

Jet pumps usually can be run in the same downhole cavity or downhole assembly as the piston pumps. This allows jet pumps to be used initially to clean up sand or trash in a well. Once the well is clean, the jet pump can easily be reverse-circulated out of the hole and replaced with a higher efficiency piston pump. If gas interference is high, then the piston pumps may require gas separation techniques to work well. Both types of hydraulic pumps can be set below the perforations without fear of possible heating problems that plague ESP motors.

9.3 THE 1 1/4-INCH JET PUMP

Recent developments in small jet pumps capable of being run on coiled tubing have increased the versatility of hydraulic pumps. Small jet pumps capable of being run inside 2 3/8- or 2 7/8-inch tubing on either 1 1/4- or 1 1/2-inch coiled tubing have been developed to eliminate the need for expensive tubing workovers. The pump bottomhole assembly can also be run-in on conventional 1 1/4-inch jointed pipe instead of coiled tubing if desired.

This small jet pump is also a "free pump," allowing the pump to be reverse-circulated to the surface for inspection or replacement. The small pump is useful for removing liquids from a gas well because it requires no workover and can easily be moved from one well to the next. The small pump can also be used for offshore applications to remove liquids from gas wells, to add artificial lift to a poor well or even as a gas lift replacement.

For offshore applications, the surface power package of the hydraulic pump can be customized to reduce the size of the footprint on the platform, where space is at a premium. In addition, the power package can be configured to supply power to several wells simultaneously, thus improving the cost efficiency while saving space.

One example of a de-watering application would be to run the coiled tubing with a rental triplex power fluid pump and operate periodically only when needed to unload the well.

Figure 9-9 shows a schematic of the Weatherford 1 1/4-inch "free" jet pump. The 1 1/4-inch pump provides an easy installation into a gas well to remove liquids. Figure 9-10 shows the small jet pump handheld.

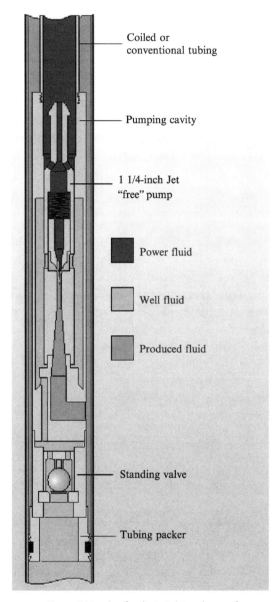

Figure 9-9. Trico (Weatherford) 1 1/4-inch jet "free" pump.

The approximate horse power for a jet pump installation is HP = 0.0006 (Depth × Gr − Pwf) (BPD), where Gr is psi/ft of production, pwf is intake pressure, psi, and BPD is total production. This neglects friction and assumes pump near perforations.

Figure 9-10. Coiled tubing "free" jet pump. (Courtesy Weatherford.)

Table 9-2
Comparative Cost to Lift Water from CBM Wells

	Beam	PCP	Jet
Initial investment (materials/rods, etc.)	$28K	$24K	$46K
Downtime/rig cost for installation/repair	$29K	$53K	$16K
Expected failures/year	3	6	2
Failure cost/year	$50K	$203K	$2K
Total first year cost	$113K	$286K	$70K
Total second year cost	$56K	$209K	$8K

9.4 SYSTEM COMPARATIVE COSTS

The following discussion summarizes one particular cost study involving beam, PCP, and jet pumps for lifting liquids off a particular coal bed methane field. The completions in this field are open-hole using the "cavitation stress completion" technique that expands large caverns in the soft coal. Most wells have 7-inch casing above the cavity. The coal is very friable, and solids production is common. Because low pressures are required for best production, pumping is needed compared to natural flow. A typical well makes 4.5 MMscf/D with 5 bbl/MMscf. With minimum detail, the findings in Table 9-2 compare using beam pumping, PCP, and jet hydraulic pumping.

The results from this table cannot be applied in general. The solids cause wear of components for all systems. Any CO_2 content can affect the elastomers in the PCPs. The PCPs must run with constant fluid level to avoid damage. Regardless, Table 9-2 shows that the jet pump systems in this case indicate higher initial investment is required but show less cost from failures in this environment, resulting in longer term more favorable economics.

These types of trends should be carefully considered when evaluating possible future pumping system evaluations. The fact that jet hydraulic pumping can be economical in some cases should not be overlooked.

9.5 HYDRAULIC PUMP CASE HISTORIES

- A 1 1/4-inch CT is used for the power fluid string. A 1-inch CT is used for the return string, and 2 7/8 inch tubing is the path for the gas flow. The HP used is 13.9 at 2600 psi power fluid for 100 bpd total fluids. The depth is 2500 ft. A free pump system was used.

- A 1 1/4-inch CT was used inside a 2 3/8 tubing from 3800 ft producing 250 to 300 bfpd.
- A 2 1/16-inch power fluid CT was used inside 3 1/2-inch tubing producing 1200 bpd with 0.950 OD jet pump producing from 5200 ft.
- A 1 1/2-inch tubing concentric to 3 1/2-inch tubing was deployed. The gas is produced up the casing annulus. The production and power fluid returns up the tubing annulus. The system produces 1800 bfpd. The pump is a 1.12-inch OD jet pump. Pump changes are possible from 4800 foot in 6 minutes.

9.6. SUMMARY

- Hydraulic pumps can be used to remove liquids from gas wells.
- A skid-mounted hydraulic pump can be used to kick off a gas well and then be moved to another well for testing, production, or longer term de-watering.
- Hydraulic pumping is generally not depth limited, and deviated or crooked wells do not present problems.
- Hydraulic reciprocating pumps can produce a low bottomhole pressure.
- A jet pump may require a fluid height over the pump of 20% of submergence.
- A jet pump is more trouble-free than a reciprocating hydraulic pump and can tolerate some solids in the production.
- Fairly high rates of more than several hundred bbls/day are possible. In general, hydraulic systems are not rate limited when removing liquids from gas wells.

REFERENCES

1. Clark, K. M., "Hydraulic Lift Systems for Low Pressure Wells," *Petroleum Engineer International*, February 1980.

2. Tait, H., "Hydraulic Pumping Systems," Southwestern Petroleum Short Course, April 21–22, 1993.

3. Martin, J. M., and Coleman, W. P., "Innovative Jet Pump Design Proves Beneficial in Coalbed Methane De-Watering Applications," Southwestern Petroleum Short Course, Lubbock, TX, April 21–22, 1993.

4. Fretwell, J. A., and Blair, E. S., "Achieving Low Producing Bottomhole Pressures in Deep Wells Using Hydraulic Reciprocating Pumps," Southwestern Petroleum Short Course, Lubbock, TX, April 21–22, 1999.

5. Simpson, D., and Deherrera, W., "Water Lifting Experiences in the CBM Fairway," Amoco presentation for Gas Well De-Watering Forum, 1999, originating from the Farmington, NM office.

CHAPTER 10

USE OF BEAM PUMPS TO DELIQUEFY GAS WELLS

10.1 INTRODUCTION

Beam pumps are the most common method used to remove liquids from gas wells. They can be used to pump liquids up the tubing and to allow gas production to flow up the casing. Their ready availability and ease of operation have promoted their use in a variety of applications; usually when the well has become so weak, other non-pumping methods cannot be supported.

Beam pump installations typically carry high costs relative to other deliquefying methods, such as foaming, plunger lift, velocity strings, or intermitting. Their initial cost can be high if a surplus unit is not available. In addition, electric costs can be high when electric motors are used to power the prime movers, and high maintenance costs often are associated with beam pumping operations. Because of the expense, alternative methods to deliquefy gas wells should be considered before installing beam pumps, or other powered pumping systems. Regardless of economics, beam pumps can work well to remove liquids from gas wells.

If beam pumps are used for gas well liquid production, the beam system often will produce smaller volumes of liquids, especially at depth. Because of the usually low volumes required to deliquefy gas wells and the fact that beam pumps do not have a "lower limit" for production and efficiency as do other pumping systems, such as ESPs, they are often used for gas well liquid production. Figure 10-1 shows an approximate depth-volume range for beam pump systems.

The presence of high gas volumes when deliquefying gas wells means that measures are often required to keep gas from entering the downhole pump and/or to allow the pump to fill and function with some gas present. Figure 10-2 shows a typical beam pumping system.

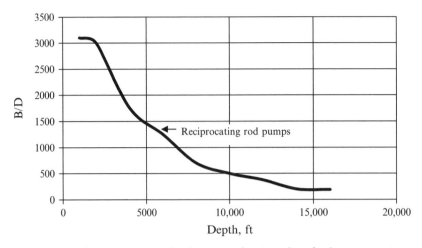

FIGURE 10-1. An approximate depth-rate application chart for beam pumping.

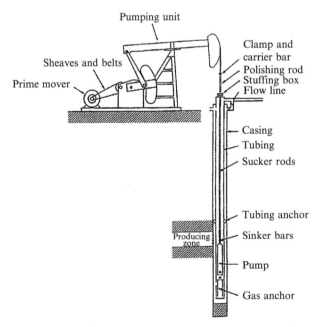

FIGURE 10-2. Schematic of beam pumping system. (Courtesy Harbison Fischer.)

This chapter discusses the primary concerns associated with the use of beam pumps to de-liquefy gas wells. Some concerns include:

- Pump-off control of the pumping system to avoid effects of over-pumping since liquid removed from gas wells is often at low rates.
- Gas separation when necessary.
- Special pumps to handle gas-induced problems, if gas enters the down-hole pump after first trying to separate gas.
- The possible use of injection systems to inject water below a packer in a water zone so gas can flow upward more easily.

10.2 BASICS OF BEAM PUMP OPERATION

The beam pumping unit changes rotary motion from the prime mover into reciprocating motion. If the prime mover is electric, it usually is a motor with a synchronous speed of 1200 RPM. Under load, it might be rotating at possibly 1140 average RPM. A beam pump is a high-efficiency device that makes good use of input electrical energy. A formula for the efficiency of a beam pump unit, or any other pumping system with input kW, is:

$$\eta = \frac{.00000736QH\gamma}{kW/.736} \tag{10-1}$$

where $\eta =$ the overall electrical efficiency of the pumping unit
$Q =$ the production from the installation, bpd
$H =$ the vertical lift of the fluid from approximately the fluid level in the casing to the surface, ft
$kW =$ the electrical power input to the motor at the surface, kilowatts
$\gamma =$ the specific gravity of the fluids being pumped

Equation 10-1 can be used for PCPs, ESPs, hydraulic pumping units, and other pumping systems. This formula, however, cannot directly be used for gaslift. Typically, PCPs and some hydraulic systems may have better efficiency than beam pump systems, and ESP systems are usually less.

A good beam pump installation can have an efficiency of more than 50%. For gas wells, however, the gas interference into the pump downhole may reduce the overall power efficiency to much less.

To reciprocate the sucker rods and pump, the high-rpm motor speed must be reduced to the required strokes per minute (SPM) for the pump.

The speed of the motor is reduced by the motor sheave and the gearbox sheave and a gearbox speed reduction of usually 30:1. The pump SPM is calculated from

$$\text{SPM} = \text{Motor RPM} \times \frac{\text{motor sheave diameter}}{\text{gearbox sheave diameter}} \times \frac{1}{\text{gearbox ratio}} \quad (10\text{-}2)$$

As an example, a beam unit with a motor sheave of 12-inch and a gearbox sheave of 37-inch, then the speed reduction from the motor to the rods or horses head will give a SPM of:

$$\text{SPM} = 1140 \frac{12}{37} \frac{1}{30} = 12.3 \text{ SPM}$$

The belts in the sheaves carry the power from the motor to the gearbox. The gearbox slows the rpm by approximately 30:1 and increases the torque to the out-put shaft of the gearbox by $\approx 30\!:\!1$, discounting some inefficiencies. The crank turned by the slow-speed shaft of the gearbox rotates and, through a pitman arm connected to the crank, moves the back end of the long walking beam up and down. The up and down motion is translated to the front of the walking beam and to the rods to impart reciprocating motion. Counterweights on the crank arm balance one-half of the fluid load and the weight of the rods in fluid to the up and downstroke loads.

The rods are connected to a polish rod at surface to pass through the stuffing box to seal the well. The polish rod is clamped on the top of the carrier bar hanging by two cables from the horse's head end of the walking beam. The rods, which are most commanly connected with couplings, are connected all the way from surface to the pump near the perforations. The rods are usually 25 feet in length (30 feet in California) and come in different grades of metallurgy. The rod string usually has a section of larger rods at the top and one or more sections of smaller diameter rods to the pump. The pump strokes up and down with motion imparted by the rods to affect a downhole stroke usually less than the surface stroke length caused by rod stretch. The formula for volumetric displacement through the pump is:

$$\text{BPD} = .1165 \, D^2 \, L \, \text{SPM} \quad (10\text{-}3)$$

where D = diameter of downhole pump, in
L = downhole stroke length at the pump, in
SPM = reciprocating cycles per minute

The downhole pump consists usually of a plunger connected to the rods with a traveling valve on the bottom of the oscillating plunger. The barrel has a standing valve on the bottom. The pump is connected to the tubing end by a top or bottom hold-down for insertable pumps (pumps that can be removed by the rods), whereas tubing pumps have the barrel screwed into the bottom of the tubing.

The pump works much better if free gas is kept from the intake of the pump. This is best accomplished by setting the pump below the pay zone or extending a short tubing section below the perforations to the pump intake. Special gas separators can be used if the pump must be set above the pay. If neither of these options is successful, special pumps are available to better handle gas.

Sand and scale can stick the bottomhole pump plunger and must be accounted for in solids-producing wells by using special pumps or filters.

10.3 PUMP-OFF CONTROL

If a beam pump is used to de-water a gas well, then often small amounts of liquid must also be produced to allow the gas to flow. The usual procedure is to pump liquids up the tubing and allow gas to flow up the casing. Because small rates of liquids may be produced, the beam system may pump at a rate higher than the well can deliver liquids over time. When a beam pump is operated at a rate beyond the capacity of the reservoir to produce liquids, the liquid level in the well is pumped below the pump intake and the pump is said to "pump-off."

Considerable literature[1,2] exists concerning beam pump systems on "pump-off" control. With gas in the pump barrel, the pump plunger initially compresses the gas on the downstroke of the pump before contacting the liquid. If sufficient gas is allowed into the barrel, the plunger can contact the fluid, causing "fluid pound" with sufficient force to ultimately damage the pump and rod string. This is of primary concern in gas wells because of the relatively high volumes of gas produced with typical low volumes of liquid.

The pump-off controller lets the beam pump operate with sufficient liquid levels to prevent damage while operating the pump at a high efficiency. The controller stops the pump when the well has been pumped off. However, some pumping systems often are allowed to operate in the pumped-off condition with continual gas interference at the pump. This results in poor efficiency and can result in "fluid pound" as the plunger contacts the fluid in a gas/liquid-filled barrel on the downstroke. Fluid pound can cause mechanical damage to the system.

10.3.1 Design Rate with Pump-Off Control

The beam pump system should be designed to be able to pump the fluid level in the annulus down to the minimum value consistent with efficient pump operation and prevention of fluid pound.

To achieve this design objective, the pump should be designed to pump at a rate given by:

$$\text{Design Rate} = \frac{\text{Maximum Inflow Capacity} \times 24 \text{ hrs/day}}{\text{Pump Volumetric Efficiency} \times \text{hrs pumped/day}}$$

(10-4)

The pump volumetric efficiency is essentially the percentage fillage of liquids in the pump barrel. For effective pump-off control, 20 hours/day pumping time is a good rule of thumb. The maximum reservoir inflow capacity should be used for the desired daily rate. Example 10-1 illustrates this equation.

Example 10.1: Design System Pumping Rate for POC (Pump-Off Control)

A gas well with maximum liquid flow capacity of 300 bfpd is to be put on beam lift to pump-off the liquids. For what rate should the pump be designed, assuming a pump volumetric efficiency of 80%.

$$\text{Design Rate} = \frac{300 \times 24}{0.80 \times 20} \text{ bfpd} = 450 \text{ bfpd}$$

(10-5)

Using this technique, the pump is designed to operate about 20 hours/ day with an 80% volumetric efficiency. The pump-off controller will turn the well off when it reaches fluid pound conditions. The operator usually sets the downtime based on production considerations. Using a typical volumetric efficiency of 80% and 20 hours/day pumping time, a simple rule of thumb is to design the beam pump system to deliver a rate equal to 1.5 times the reservoir maximum inflow capacity. Cheaper "timers" can also be used to interrupt the "pounding" of fluid.

Design Rate = $1.5 \times$ Maximum Inflow Capacity

10.3.2 Use of Surface Indications for Pump-Off Control

An important tool for diagnosing beam pump problems is the surface dynamometer card or load (in the rods) vs. position at the top of the rod string. Usually the shape of the surface dynamometer card is used by a computerized system to determine when the well is beginning to "pump-off." Other systems include using a calculated downhole dynamometer card shape, cycle time, vibration, and other techniques to determine pump-off.

The surface dynamometer card is a plot of load and position on the polished rod just above the top rod. The shape of the surface card and especially the calculated card for the load and position in the rod just above the pump can be indicative of problems or good operation at the downhole card.

Figure 10-3 shows that the surface card can indicate the condition of the pump. The two top cards show the surface dynamometer cards. The surface card is shown for a full pump, and the pump card below it shows that the pump card is almost rectangular, which indicates a full pump with few or no problems.

For the two right-most cards, however, the bottom pump card shows that the bottom right of the pump card is not outlined by a load-position

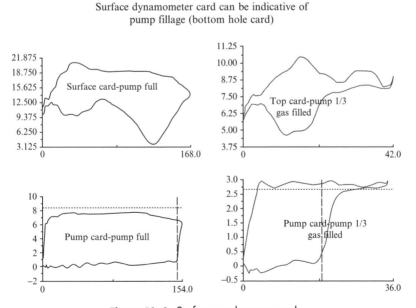

Figure 10-3. Surface and pump cards.

line. This is because the TV (traveling valve load) has not been released from the rods until about one-third into the downstroke. This is because the gas and/or fluid below the TV has not accumulated enough pressure below the TV to open it and drop the load on the TV and the rods. The load can be dropped gradually (gas interference) or quickly when the TV hits the fluid. The worst situation is for the TV to hit fluid for the first time somewhere near the middle of the downhole stroke when the plunger is traveling much faster than at the beginning and end of the stroke.

Because, as shown above, the surface card can indicate what is happening at the pump, then pump-off control can use the surface dynamometer card to control on since it can indicate when the bottomhole pump card is full and when it is beginning to fill with some gas.

Figure 10-4 shows a surface card with a computerized set point, indicated by the "+," that indicates the point at which the POC will shut the pumping action off and allow the well to accumulate liquids. The latter stages of lesser pump fillage pumping action are not allowed to occur.

It is usually desirable to shut-in the pumping system when the barrel becomes no less than perhaps 80% to 85% full, although conditions vary. Pump-off control is simply a method of oversizing the pumping action of the pumping system and then shutting off the system when gas interference begins at the pump. Another harder to control method that would achieve the same results for production would be to simply maintain a low fluid level over the pump at all times.

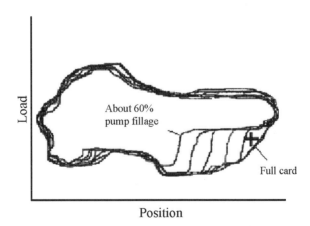

Figure 10-4. Surface dynamometer card showing various degrees of pump fillage.

10.4 GAS SEPARATION TO KEEP GAS OUT OF THE PUMP

When removing liquids from a gassy well using beam pumps, it is possible that the pump will be subjected to gas interference. Measures may be needed to separate the gas[4,5] from the liquid stream before it enters the pump to prevent "gas locking," low efficiency, reduced production, and possible damage from fluid pound.

Before outlining some guidelines, let us first identify what is meant by gas interference and fluid pound. For fluid pound the pump intake is at a low pressure, and the barrel is partially full of liquid and partially full of gas. The plunger comes down on the downstroke and passes through gas and then suddenly impacts the fluid in the barrel, causing "fluid pound." This can cause rods to unscrew, rod and tubing damage as the rods bend to the tubing, and other bad effects. This indicates that the well is being pumped off or is pumping too fast. A pump-off controller may prevent this. When pump-off control is installed or the pump is slowed, the fluid pound will cease, but gas interference caused by gas from the formation coming with the production will still occur.

For gas interference, the pump intake is usually at a higher pressure, and the pump is volumetrically filled with part gas and part fluid. The gas, or a mixture of liquids and gas, higher pressure gradually increases the pressure in the barrel below the plunger and helps cushion the impact of the plunger with the fluid on the downstroke. This still causes low pump fillage or efficiency and low production but may not cause the mechanical damage that the fluid pound causes. There is usually a fluid level above the pump for gas interference, and the gas is coming with the fluid from the perforations.

Figure 10-5 shows what the calculated bottomhole dynamometer looks like for fluid pound or for gas interference.

The following rules are based on the height of fluid over the pump in the annulus of the well. This height of fluid measured by acoustic means should be corrected for gas content in the fluid level.[4]

If the fluid level is low:

- No gas interference in the downhole pump is indicated, then "good job."
- If some gas interference present, but no fluid pound, then still acceptable.
- If gas interference with possibly damaging fluid pound present, then consider gas separation.

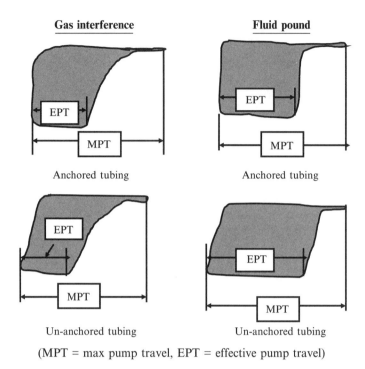

(MPT = max pump travel, EPT = effective pump travel)

Figure 10-5. Pump dynamometer card showing various degrees of pump fillage.

If the fluid level is high:

- No gas interference is present, then pump at a higher rate to lower well pressure and produce more gas up the annulus. This would be a high priority.
- Gas interference is present; consider gas separation so you can pump liquids at a higher rate and allow more gas to be produced. This would be a high priority.

The following methods are used to separate gas from the downhole pump intake.

10.4.1 Set Pump Below Perforations

One of the simplest and best methods to separate gas from the liquid at the pump is to set the pump below the perforations. The slow downward velocity (less than $\approx \frac{1}{2}$ ft/sec is needed) of the liquid in the casing-tubing annulus down to the pump intake allows the gas to separate from the liquids and migrate freely up the annulus. At the same time, the liquids migrate downward to the pump intake carrying a minimal amount of gas through the pump. If this downward velocity is less than approximately $\frac{1}{2}$ ft/sec, then the amount of gas being carried downward is minimal, especially if only water is being pumped. If the pump cannot be set below the perforations, then other types of gas separators must be considered.

10.4.2 "Poor-Boy" or Limited-Entry Gas Separator

One leading type of gas separator is the so-called "poor-boy" gas separator. Various modifications of the "poor-boy" separator have been widely used in the industry over the past 20 years. Figure 10-6 shows a schematic of the "poor-boy" separator, which is also referred as the limited entry separator.

The device is named "limited entry" because the entry for the fluid is also the entry for stray bubbles, which if entrained, have no place to escape. The poor-boy separator is designed so that the downflow in the annular area inside of the separator is less than $\frac{1}{2}$ ft/sec, so that any bubbles in the flow will not be carried into the pump intake through the dip tube. In gassy wells, however, the free gas component makes it difficult to determine when the actual velocity is below $\frac{1}{2}$ ft/sec relative to the surface production. In addition, bubbles that migrate inside the separator are unable to escape and can eventually "gas-lock" the separator.

A rough rule of thumb is that if a gassy well is producing over 200 bpd, or more, it will gas-lock a poor-boy separator.

A simple modification that has been made to the poor-boy separator, making it more applicable to gassy wells, is shown in Figure 10-7. The modification is to use a stinger to set below the perforations but not the entire pump body. This modification allows a very slow velocity down to the intake, allowing gas to come up the annulus. The inlet of the stinger is positioned below the perforations, allowing the gas to separate from the low velocity fluid in the annular region. The length of the stinger should be kept to a minimum. If the stinger is too long, then the combination of the frictional pressure drop and pressure head caused by elevation change can bring gas out of the solution in the stinger and defeat this system.

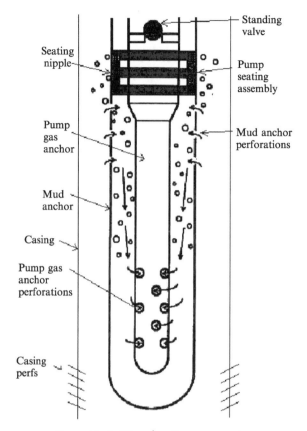

Figure 10-6. "Poor-boy" gas separator.

10.4.3 Collar-Sized Separator

Another separator is the collar-sized gas separator[4] shown in Figure 10-8. It is fairly inexpensive, has large intake and discharge ports, and can be expected to give good results at fairly low pressures (less than a few 100 psi).

The collar-sized gas separator selected should be the same size as the tubing unless the pump capacity exceeds the gas separator capacity. In this case, a larger gas separator should be selected that has a liquid capacity equal to or greater than the pump capacity. At high liquid and gas rates, even an optimum-size gas separator in limited size casing may not have the capacity to separate all of the free gas from the liquids at low pump intake pressures.[4]

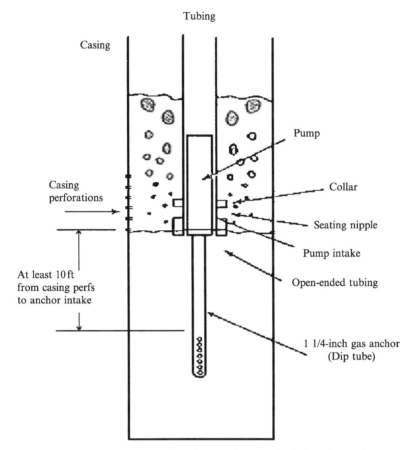

Figure 10-7. Separator using a dip-tube to allow intake below the perforations.

A beam pumping system operating in a gassy environment should have some sort of effective gas separation downhole. Although a wide variety of gas separation systems are given in the literature, those discussed here are found to be among the most successful.

10.5 HANDLING GAS THROUGH THE PUMP

If separators are not successful in eliminating gas from the pump, then special pumps or pump construction will assist in handling gas through the pump as a last resort.

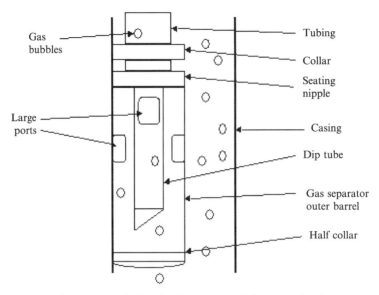

Figure 10-8. Collar-sized separator (Echometer, Inc.).

10.5.1 Compression Ratio

In some cases, the produced gas volume is so high that most of the gas cannot be separated. In this case, the pump must be designed to minimize the effects of the free gas that will enter the pump.

As discussed in Section 10.2, the traveling valve must open on the downstroke for the pump to work effectively. When pumping gas through the pump, the pump must compress the gas in the pump on the down-stroke to a pressure greater than the pressure above the traveling valve to force the traveling valve off its seat. If the traveling valve does not open, the pump action will continue but the pump cannot pump liquid. This condition is called "gas-locking."

Beam pump installations can be designed so that they are not suscep-tible to gas-lock regardless of the amount of gas passing through the pump.[6] If the compression ratio of the downhole pump is high enough to always push fluids above a valve and into the tubing it will not gas-lock even if it contains 100% gas. This certainly will not improve the volumetric efficiency of the pump, but the pump will not gas-lock.

The compression of the pump discussed in this section is how much the fluid below the traveling valve is compressed on the downstroke. This pressure should be sufficient to accumulate enough pressure to open

Compression ratio

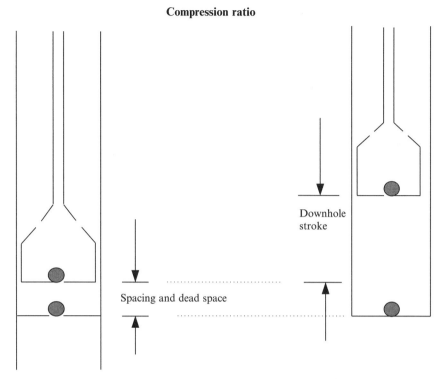

Figure 10-9. Beam pump compression ratio.[6]

the traveling valve on the downstroke. If the traveling valve always opens on the downstroke, then the pump will not gas-lock.

The definition of compression ratio (CR) is given by (Figure 10-9):

$$CR = \frac{\text{Downhole Stroke} + \text{Spacing Clearance} + \text{Dead Space}}{\text{Spacing Clearance} + \text{Dead Space}} \quad (10\text{-}6)$$

The key to attaining a high compression ratio is to maximize the downhole stroke at the pump while minimizing the spacing clearance and dead space. Typically, expensive surface unit changes or adjustments are needed to increase the stroke, but careful spacing can drastically increase the compression ratio.

The pull rod should be cut in the shop so that the clearance between the traveling valve and the standing valve is less than approximately $^{1}/_{2}$ inch when the pump is at its down-most position. In the well, the pump must be spaced to a bare minimum, taking care that the pump does not "tag" or strike the bottom on the downstroke, but the standing valve assembly must come close to the traveling assembly on the downstroke to minimize the dead space in the pump.

If pump spacing and pull rod length are considered, many pump gas handling problems will be solved. The pull rod length is easily overlooked because you cannot see how long the pull rod is until you disassemble the pump.

10.5.2 Variable Slippage Pump to Prevent Gas-Lock

The H-F variable slippage pump shown in Figure 10-10 is primarily for gas-locking conditions. This pump has eliminated gas-lock in each field test to date.

Leakage is allowed to occur from over the plunger to under the plunger at the end of the upstroke due to a widened or tapered barrel. Although this reduces pump efficiency, enough liquid allowed to leak below the plunger ensures that the traveling valve opens on the downstroke and that gas-lock will not occur.

10.5.3 Pump Compression with Dual Chambers

The pump in Figure 10-11 works by holding back the hydrostatic pressure in the tubing on the downstroke while still allowing fluid and gas to enter the upper chamber. The fluid is compressed once on the downstroke into an upper smaller chamber. It is then compressed on the upstroke into the tubing. If the upper and lower compression ratios are 20:1, then the overall compression ratio is 400:1.

10.5.4 Pumps That Open the Traveling Valve Mechanically

Several pumps have a mechanism to automatically open the traveling valve on the downstroke, thereby preventing gas-lock. Some have sliding mechanisms, whereas others have devices that directly dislodge the traveling valve from its seat if not already dislodged by pressure. The pump assembly in Figure 10-12 uses a rod to force the traveling valve ball off the seat on the downstroke.

10.5.5 Pumps to Take the Fluid Load Off the Traveling Valve

Figure 10-13 shows a slide above the pump that seals the pressure above the pump from being on the top of the traveling valve on the downstroke. There are other pumps that use this concept.

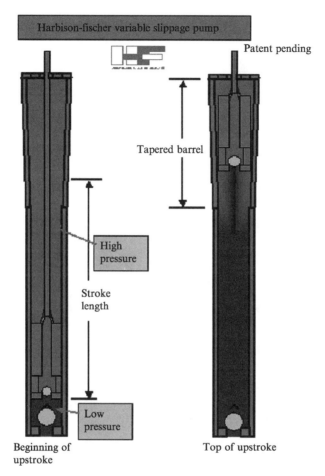

Figure 10-10. Example of a pump that uses designed leakage to prevent gas-lock. (Courtesy Harbison-Fischer.)

There are many other specialty pumps. First, try to separate the gas using completion techniques with the pump below the perforations. If this fails, try a gas separator. If this still fails, then try the more exotic pumps to handle gas.

10.6 INJECT LIQUIDS BELOW A PACKER

In recent years, methods have been developed to separate the liquid and gas phases downhole and then re-inject the liquids (must be water) back into the formation below a packer. This eliminates both the need for

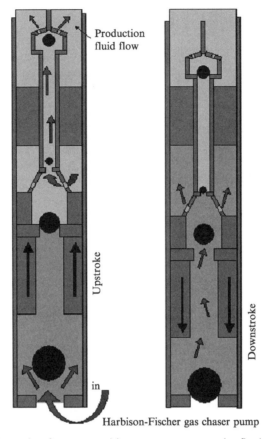

Harbison-Fischer gas chaser pump

Figure 10-11. Example of a pump adding compression to the fluid on the upstroke (Harbison-Fischer).

disposal of the water at the surface and the means required to lift the liquids to the surface. Once the liquids are re-injected, the gas can flow freely up the casing-tubing annulus. Several commercial devices are available[7–9] to do this. The concept shown in Figure 10-14 uses gravity as both the separation mechanism and the injection mechanism. The water is pumped up the tubing. The bypass seating nipple allows water pressure and flow to bypass the pump. The pressure exerted on the formation below the pump injects the water. The higher the fluid column in the tubing generated by the pump, the greater the pressure on the formation. If a larger pressure is needed than provided by a full column of liquid in the tubing, then a backpressure regulator can be placed on the surface of the tubing. Cases requiring 300 psi surface pressure and greater are reported to inject at the desired rate.

Hart gas lock breaker
standing valve assembly

Figure 10-12. Example of a pump that mechanically opens the traveling valve on the downstroke with a rod that lifts the traveling ball off its seat.

10.7 OTHER PROBLEMS INDICATED BY THE SHAPE OF THE PUMP CARD

In the previous materials, figures were shown of the surface dynamometer cards and the bottomhole pump cards and how the shape of the cards can be used to help diagnose gas problems, such as fluid pound and gas interference.

However, other problems can also be diagnosed by the shape of the cards, as shown in Figure 10-15.

Unanchored tubing

The top two cards show a pump that is full of liquid; however, the one to the right has the tubing unanchored. This allows the tubing to travel upward some on the upstroke, and the pickup of the load takes place over a distance of upstroke, resulting in a slanting of the sides of the card with the un-anchored tubing.

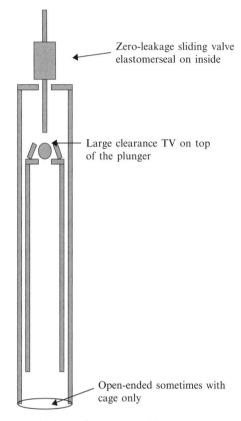

Zero-leakage sliding valve
elastomerseal on inside

Large clearance TV on top
of the plunger

Open-ended sometimes with
cage only

Figure 10-13. Quinn multiphase flow pump: Slide above pump closes on down-stroke to take fluid load off of the valve. For usual pump with TV and SV, the load off the TV will allow the TV to open with gas and liquids below the pump and reduce fluid pound (Quinn Pumps, Canada).

Leaky traveling valve

The second row of two cards illustrates a leaking traveling valve (TV). The cards with the leaky TV show a rounding of the top of the card. The load is not immediately picked up, and as the top of the upstroke is neared, the load begins to fall off again. This results in low or no production.

Leaky standing valve

The third row of two cards illustrates a leaking standing valve (SV). The cards for the leaky SV show a rounding of the bottom of the card. The leaky SV lets fluid out below the TV, delaying the loss of load and

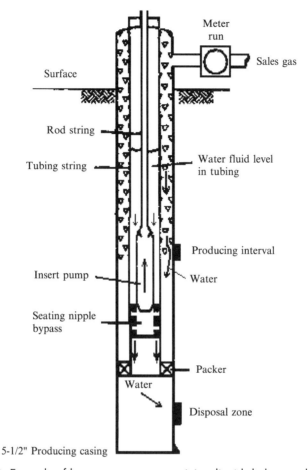

Figure 10-14. Example of beam pump system to inject liquids below packer so that gas can flow unobstructed (Harbison Fischer bypass seating nipple).

opening the TV on the rods above the pump. At the end of the downstroke, the loss of fluid allows the TV to close prematurely, bringing up the load again on the rods prematurely.

Improper spacing

The fourth row of cards shows improper pump spacing. The pump can be spaced too high or too low, resulting in the pump "tagging" on the upstroke (spaced too high) or on the downstroke (spaced too low). Although this can cause rod and pump damage, it is good practice to space the pump as low as possible without tagging the pump in a gassy environment.

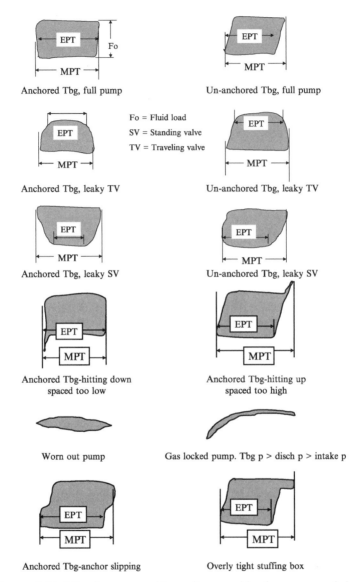

Figure 10-15. Miscellaneous problems diagnosed by the pump card shape.

Worn pump

The worn-out pump is a combination of the leaky TV and leaky SV. It may be producing no fluid at the surface.

Gas-locked pump

The gas-locked pump is in a situation such that on the downstroke, the TV never opens, and no fluid is being pumped. As fluid in the annulus accumulates, the gas-lock will clear. Spacing the pump as low as possible without tagging will usually prevent this from occurring for all but the gassiest wells. A long stroke will also help prevent gas-lock.

Tubing anchor slipping

The card shown for a jagged pickup and release of the load is for the case of the tubing anchor slipping. This is because although the anchor is placed and set, it is slipping on the upstroke and the downstroke, leading to an erratic load pickup and release. This situation is bad because the slipping anchor can lead to casing wear and rod and tubing wear and usually indicates a need to pull and reseat the anchor at a slightly different location.

Tight stuffing box

The card for the overly tight stuffing box (last card in this figure) should show an extra amount of fluid load (Pabove-Pbelow × Area of Pump) on the card, which is exhibited as an extra vertical thickness for the card. The extra friction is usually released at the top of the stroke, leading to the shape of the card.

There are many other problems associated with beam pumps, including leaky tubing, slipping belts, worn sheaves, worn gearboxes, improperly sized motors or prime movers, incorrect design, such as pumping too fast with a small diameter pump, and other factors too numerous to mention. Many factors dealing with gas have been mentioned in this chapter because gas problems at the pump are often a concern when trying to pump liquids off of gas wells.

10.8 SUMMARY

Although beam pumps are often used for de-watering a gas well, special methods may be required to prevent gas interference.

- Gas interference is most often and easily handled by setting the pump below the perforations and flowing the gas up the annulus.

- A well-spaced pump with a maximum pull rod length solves many gas-lock problems.
- If the pump does not fit below the perforations for gas separation, then separators or, as a last resort, specialty pumps may be required to combat pump gas interference.
- To handle produced water, the beam pump system may be incorporated with a system to inject water below a packer to a water zone. This method eliminates water hauling charges and leaves a free path for gas to flow to the surface.
- For less than 100 BPD, consider use of 1.06-inch pump, 5/8-inch rods to 7000 ft, possibly using no anchor, and use pump-off control.[10]

REFERENCES

1. Lea, J. F., "New Pump-Off Controls Improve Performance," *Petroleum Engineer International*, December 1986, pp. 41–44.

2. Neely, A. B., "Experience with Pump-off Control in the Permian Basin," SPE 14345, presented at the Annual Technical Conference and Exhibition of the SPE, Las Vegas, NV, September, 22–25, 1985.

3. Dunham, C. L., "Supervisory Control of Beam Pumping Wells," SPE 16216, presented at the Production Operations Symposium, Oklahoma City, OK, March 8–10, 1987.

4. McCoy, J. N., and Podio, A. L., "Improved Downhole Gas Separators," Southwestern Petroleum Short Course, Lubbock, TX, April 7–8, 1998.

5. Clegg, J. D., "Another Look at Gas Anchors," Proceedings of 36th Annual Meeting of the Southwestern Petroleum Short Course, Lubbock, TX, April 1989.

6. Parker, R. M., "How to Prevent Gas-Locked Sucker Rod Pumps," *World Oil*, June 1992, pp. 47–50.

7. Enviro-Tech Tools Inc. brochure on the DHI (Down Hole Injection) tool.

8. Grubb, A., and Duvall, D. K., "Disposal Tool Technology Extends Gas Well Life and Enhances Profits," SPE 24796, presented at the 67th Annual SPE Conference in Washington DC, October 4–7, 1992.

9. Williams, R., et al., "Gas Well Liquids Injection using Beam Lift Systems," Southwestern Petroleum Short Course, Lubbock, TX, April 2–3, 1997.

10. Elmer, W., and Gray, A., "Design Considerations when Rod Pumping Gas Wells," First Conference of Gas Well De-Watering, SWPSC/ALRDC, March 3–5, 2003, Denver.

CHAPTER 11

GAS LIFT

11.1 INTRODUCTION

Gas lift is an artificial lift method whereby external gas is injected into the producing flowstream at some depth in the wellbore. The additional gas augments the formation gas and reduces the flowing bottomhole pressure, thereby increasing the inflow of produced fluids. For de-watering gas wells, the volume of injected gas may be designed so that the combined formation and injected gas will be above the critical rate for the wellbore.[1]

Although gas lift may not lower the flowing pressure as much as an optimized pumping system, several advantages make gas lift the artificial lift method of choice.

Of all artificial lift methods, gas lift most closely resembles natural flow and has long been recognized as one of the most versatile artificial lift methods. Because of its versatility, gas lift is a good candidate for removing liquids from gas wells under certain conditions. Figure 11-1 shows the approximate depth-pressure ranges for application of gas lift.

The most important advantages of gas lift over pumping lift methods are:

- Most pumping systems become inefficient when the GLR exceeds some high value, typically approximately 500 scf/bbl ($90m^3/m^3$), because of severe gas interference. Although remedial measures are possible for conventional lift systems, gas lift systems can be directly applied to high GLR wells because the high formation GLR reduces the need for additional gas to lower the formation-flowing pressure.

- Production of solids will reduce the life of any device that is placed in the produced fluid flowstream, such as a rod pump or ESP. Gas lift systems generally are not susceptible to erosion caused by sand production and can handle a higher solids production than conventional pumping systems.
- For some applications, a higher pressure gas supply of a gas zone may be used to auto-gas-lift another zone.
- In highly deviated wells, it is difficult to use some pumping systems because of the potential for mechanical damage to deploying electric cables or rod and tubing wear for beam pumps. Gas lift systems can be used in deviated wells without mechanical problems. However, gas injected in near-horizontal sections will not reduce gravity pressure effects and will in fact increase frictional losses.

Gas lift is the best alternative artificial lift method that addresses all of these shortcomings.

Another advantage that gas lift has over other types of artificial lift is its adaptability to changes in reservoir conditions. It is a relatively simple matter to alter a gas lift design to account for reservoir decline or an increase in fluid (water) production that generally occurs in the latter stages of the life of the field. Changes to the gas lift installation can be made from the surface without pulling tubing by replacing the gas lift valves

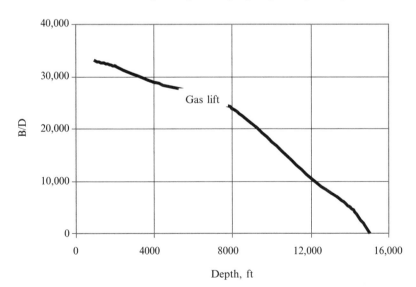

Figure 11-1. An approximate depth-rate feasibility chart for conventional continuous gas lift.

via wireline and reusing the original downhole components. Gas lift is limited in how low the producing bottomhole pressure can be achieved.

The two fundamental types of gas lift used in the industry today are "continuous flow" and "intermittent flow."

11.2 CONTINUOUS GAS LIFT

11.2.1 Basic Principles of Continuous Gas Lift

In continuous-flow gas lift, a stream of relatively high pressure gas is injected continuously into the produced fluid column through a downhole valve or orifice. The injected gas mixes with any formation gas to lift the fluid to the surface by one or more of the following processes:

- Reduction of the fluid density and the column weight so that the pressure differential between the reservoir and the wellbore will be increased.
- Expansion of the injected gas so that it pushes liquid ahead of it, which further reduces the column weight and increases the differential between the reservoir and the wellbore.
- Displacement of liquid slugs by large bubbles of gas acting as pistons.

11.3 INTERMITTENT GAS LIFT

As the bottomhole pressure declines in gas wells, a point is reached when the well is not as economic with continuous gas lift, and the well is converted to intermittent gas lift. This point may be around 200 bbls/day. This conversion can also use the identical downhole equipment (mainly the gas lift valve mandrels), yet fully adapt the well to intermittent flow. In this case, the unloading valves are replaced with dummy valves to block the holes in the mandrels and prevent injection gas from passing into the production stream. The operating valve is then replaced with a production pressure valve with a newly set pressure capacity reflecting the desired fluid level to be reached in the tubing before the well is lifted. Some intermittent designs have valves above the bottom value to add gas as a liquid slug passes.

Fitting the operating valve with the largest possible orifice will greatly improve the efficiency in an intermittent gas lift system. The large orifice diameter exerts a minimum restriction to the flow of the injection gas. The injection gas will then quickly fill the tubing below the fluid,

Table 11-1
Maximum Flow Conditions for
Intermittent Lift

Tubing Size (inch)	Maximim Flow rate for Intermittent Lift
2 3/8	150 bpd
2 7/8	250 bpd
3 1/2	300 bpd
4 1/2	Not Recommended

ultimately lifting the "slug" of liquid to the surface with the minimum amount of lift gas.

The optimum time to convert a gas lift well from continuous lift to intermittent lift is a function of the reservoir pressure, the tubing size, the GLR, and the flow rate of the well. Although the individual well conditions will dictate the optimum time for conversion, Table 11-1 lists some good rules of thumb to use to estimate the best time to convert to intermittent lift.

It is becoming common practice to use a plunger (Chapter 7) to increase the production from wells on intermittent lift. The lift gas is injected below the plunger, and the plunger acts as a physical barrier between the lift gas and the fluid to reduce the fluid fallback around the gas slug that is characteristic of intermittent lift operations. The plunger extends the lift of the well by more effectively removing water from the formation. A plunger with extensions can be used so that it can pass by gas lift mandrels if needed as discussed below. A rule of thumb is if the slug of liquid being produced travels at > 1000 ft/min, a plunger is not needed. However, if the well is being intermitted and the slug of liquid reaches the surface with a velocity of < 1000 ft/min, then using a plunger will be advantageous to reduce liquid fall-back.

11.4 GAS LIFT SYSTEM COMPONENTS

Figure 11-2 shows what a typical continuous gas lift system includes:

• Gas source
• Surface injection system, including all related piping, compressors, control valves, etc.
• Producing well completed with downhole gas lift equipment (valves and mandrels)
• Surface processing system, including all related piping, separators, control valves, etc.

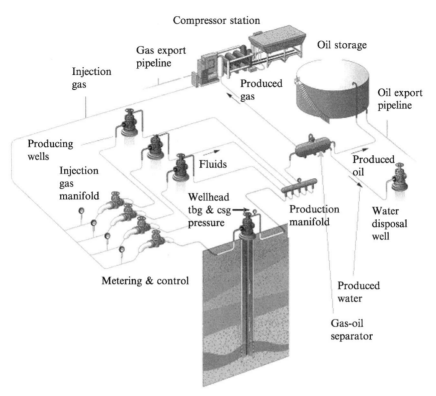

Figure 11-2. Continuous gas lift system (Courtesy of Schlumberger-Camco).

The gas source is often reservoir gas produced from adjoining wells that has been separated, compressed, and re-injected. A secondary source of gas may be required to supply any shortfall in the gas from the separator. The gas is compressed to the design pressure and is injected into the well through the gas lift operating valve, where it enters the tubing string at a predetermined depth.

For conventional gas lift, valves or orifices should be used to port gas to the tubing, rather than holes or simply the end of the tubing string, so that the gas stream is well dispersed in the liquid column and flows smoothly.

"Gas cycling" is a method to flow additional gas down the annulus and into the bottom of the tubing.[2] This is possible because the amount of gas is high in the tubing relative to the fluids so that severe slugging does not occur as it would with a lower operational GLR as would be typical for gas lifting oil wells.

11.5 CONTINUOUS GAS LIFT DESIGN OBJECTIVES

Gas lift increases well production by primarily reducing the density of the produced fluid, thereby decreasing the flowing bottomhole pressure. Some reduction in pressure is achieved by a "scrubbing" action of the gas bubbles. This is accomplished by introducing the injection gas at an optimum (usually maximum) depth, pressure, and rate into the produced fluid stream. The valve through which the gas is injected into the wellbore fluid stream under normal operating conditions is called the "operating valve."

Nodal Analysis (Chapter 2) can be used to evaluate several tubing sizes and GLRs to determine possible production increases for the different tubing sizes and GLRs. The well becomes a candidate for gas lift when the artificially increased GLR significantly increases the well production. Another way to think of gas lift for gas wells is to inject a sufficient additional volume of gas to keep the total gas velocity (from produced + injected gas) above the critical velocity for the well. If the gas velocity is always above critical, then liquid loading will never occur.

The efficiency of a gas lift system is highly influenced by the depth of the operating valve. As the depth of the operating valve is increased, more and more of the hydrostatic pressure of the heavier fluid (and gas) column is taken off of the formation, reducing the bottomhole pressure and increasing production. Typically, before a gas lift well is brought online, it is filled or partially filled with kill fluid from the workover operation. To bring the well into production, the well must first be "unloaded" by injecting high-pressure gas into the annulus to displace the kill fluid in the annulus down to the operating valve.

An extremely high surface injection pressure is needed to push the liquid level to the depth of the operating valve. In most installations, this high injection pressure is not available. Several gas lift valves are required to allow the available surface pressure to feed gas to the well at increasing depths until the operating valve at maximum depth is reached. This process is called "unloading the well," and the additional upper valves are called "unloading valves."

Unloading valves are placed at various depths and have different opening/closing pressures to step the injection gas down to the design injection depth. These unloading valves are designed to have a particular port size and set to specific opening pressures to allow the annular fluid level to pass from one valve to the next. The design of the gas lift system includes the size, pressure rating, depth and spacing of the unloading valves, the optimum depth of the operating valve to maximize recovery, the size of the operating valve orifice, and the injection rate and pressure of the lift gas.

The correct spacing of the unloading valves is critical. Valves spaced too far apart for the injection parameters will not allow the well to completely unload. In this case, injection gas will enter the production stream too high in the well, significantly lowering the system efficiency and, more importantly, then well's production.

Determining the best gas lift design requires considerable knowledge of the well conditions, both present and future. These calculations usually are performed by sophisticated commercial software packages or design charts supplied by gas valve manufacturers. The complete fundamentals of gas lift design and optimization are beyond the scope of this text although field applications[3] of gas lift technology for gas wells are presented in this chapter.

11.6 GAS LIFT VALVES

Gas lift valves fall into one of three major categories:

- Orifice valves
- Injection pressure-operated (IPO) valves
- Production pressure-operated (PPO) valves

Schematic examples of injection and production pressure operated valves are given in Figure 11-3. By far, the type 1 and possibly type 2 valves are used to de-water lower pressure gas wells.

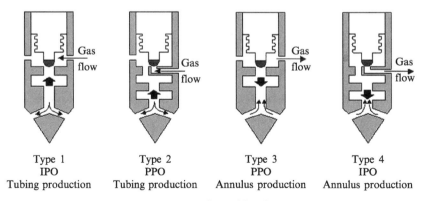

Figure 11-3. Typical gas lift valve types.

11.6.1 Orifice Valves

Strictly speaking, orifice valves are not valves because they do not open and close. Orifice valves contain orifices or holes that provide a communicating port from the casing to the tubing. Because they do not actually function as valves, orifice valves are used only as operating valves to provide the correct injection flow area as required by the valve design and to disperse properly the injected gas to minimize the formation of slugs. Orifice valves are typically used only for continuous flow applications. The valve assembly includes a check to prevent tubing to casing flow, when producing up the tubing.

11.6.2 IPO Valves

IPO (sometimes called casing pressure operated or injection pressure operated) valves are the most common valves used in the industry to unload gas lift wells. Although somewhat influenced by the pressure of the flowing production fluid, IPO valves are controlled primarily by the pressure of the injection gas.

Figure 11-4 shows a schematic of an IPO gas lift valve where the injection pressure is applied to the base of the bellows, and the produced fluid pressure is applied to the ball (stem tip) through the valve orifice area. Because the bellows area is much larger than the orifice area, the injection pressure dominates control of the valve operation.

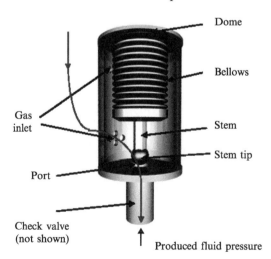

Figure 11-4. Schematic of gas lift valve.

Injection pressure valves act like backpressure regulators and close when the backpressure (casing pressure) reaches a predesignated "minimum" value. This minimum value is designed to be when the kill fluid in the casing/tubing annulus, being pushed downward by the injection gas during the unloading process, just reaches the next lower valve. This allows the upper valve to close to the flow of injection gas, allowing the pressure to continue to push the fluid level further down the annulus and allowing gas to eventually reach the operating valve.

11.6.3 PPO Valves

PPO valves (sometimes called tubing pressure valves) are primarily operated by changes in pressure of the production fluid. Unloading is then controlled primarily by the reduction in hydrostatic pressure in the production stream by injecting lift gas.

PPO valves are used typically for:

- Production fluid is produced through the annulus
- Dual completions where two gas lift systems are installed in the same well to produce two differently pressured zones
- Intermittent lift

However, some effectors use PPO valves for normal tubing flow installations, sometimes because they may unload to a deeper depth with the same injection pressure. Use of PPO valves can lead to instability in certain situations.

PPO valves are ideal for intermittent lift applications because the valve is designed to remain closed until a sufficient fluid load is present in the tubing; at which time the valve opens, producing the liquid.

Once an unloading valve closes during the unloading process, it should remain closed. Both injection and production pressure valves use a charged bellows (typically pressurized with nitrogen), a spring, or sometimes both to obtain the valve closing force. The nitrogen charged bellows is the most common. The bellows type valves are set to the design pressure in a controlled laboratory environment by the valve shop.

All gas lift valves are equipped with reverse flow check valves to prevent backflow of fluid through the valve. For sub-sea completions, where minimal intervention is a design objective, the spring-loaded valve may provide the most reliability, since in the event of a bellows rupture the spring will keep the stem on seat and the valve will remain closed. The spring-loaded valve is also not sensitive to temperature variations as is the nitrogen-charged bellows.

11.7 GAS LIFT COMPLETIONS

The heart of a gas lift installation is the gas lift valves. Their placement in the tubing string is fixed during the installation of the tubing by the gas lift mandrels. Gas lift mandrels are placed in the tubing string to position each gas lift valve to the desired depth.

Two basic "conventional" gas lift systems are used today: (1) systems using conventional mandrels with threaded non-retrievable gas lift valves and (2) systems using side pocket mandrels (SPM) with retrievable gas lift valves.

Conventional mandrels accept threaded gas lift valves mounted on the outside of the mandrel. These valves can only be retrieved and changed by pulling the tubing and are usually not run where workover costs are high.

SPMs allow the gas lift valves to be retrieved using slickline from the surface without the need to pull the tubing. These mandrels are most commonly used today off-shore. Both systems are discussed later.

11.7.1 Conventional Gas Lift Design

A schematic of a gas lift system using conventional mandrels is shown in Figure 11-5. With this system, gas lift mandrels and valves are installed at the surface when the tubing is run in the well. The valves are threaded into the mandrels and therefore cannot be removed without removing the entire tubing string. Gas lift designs using conventional mandrels are among the lowest cost gas lift designs available. However, they should be used where tubing pulling costs are low such as shallow on-shore installations.

An added benefit of using conventional mandrels, particularly when removing liquids from gas wells, is that they can readily integrate with plunger lift systems. This is not the case for installations using SPMs. The ID of a conventional mandrel is relatively uniform, but the internal pocket of a side pocket mandrel is eccentric to permit the insertion of gas lift valves via slickline. This presents a problem for plunger operations because as the plunger assembly enters the SPM's eccentric pocket it allows gas to bypass liquid. This typically results in a loss of plunger velocity and, in some cases, makes it difficult for the plunger to reach the surface. Some operators have successfully adapted extensions to the

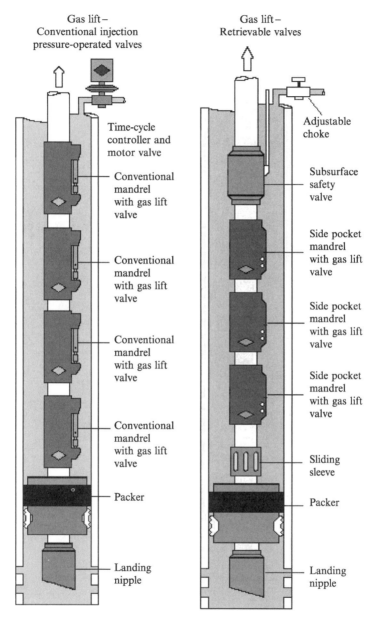

Figure 11-5. Gas lift design using conventional mandrels *(left)* and side pocket mandrels with wireline retrievable valves *(right)*. (Courtesy Schlumberger-Camco.)

plunger to effectively straddle the pockets, however, these have succeeded only for shallower wells. The long plunger with extensions requires a much taller wellhead.

SPMs were developed to reduce the costs of changing a gas lift system to maintain a gas lift valve design that optimizes production as well conditions change. A schematic of a SPM is shown in Figure 11-6. The primary feature of SPMs is the internally offset pocket, which accepts a slickline-retrievable gas lift valve. The pocket is accessible from within the tubing using a positioning or kickover to place and retrieve the valves. The gas lift valves use locking devices that lock into mating recesses in the SPM. Although both conventional and SPM mandrels are installed in the well in the same manner, only the SPM system is serviceable with slickline operations for postcompletion repair or well maintenance.

The high-pressure gas in a gas lift system is usually supplied by a central compressor that compresses the gas produced by the field for re-injection into those wells on gas lift. If the field gas supply is insufficient to meet the needs of the artificial lift system, more gas is generally obtained from the sales line.

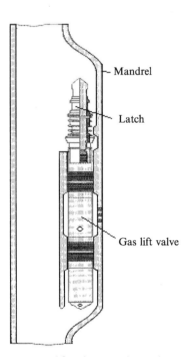

Figure 11-6. Gas lift valve in side pocket mandrel.

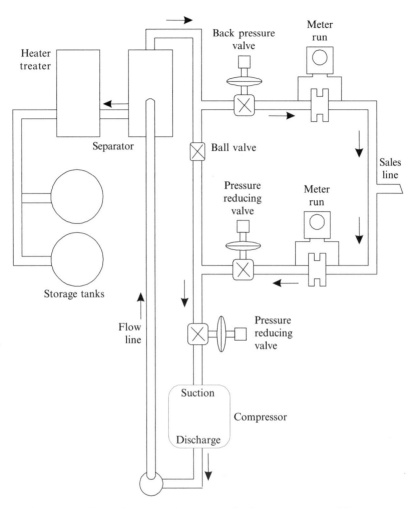

Figure 11-7. Typical compression system for low-pressure gas lift system.

Gas lift compression can also be supplied for individual wells when one or two wells in a field are being lifted with gas lift. Small well-site compressors are typically skid mounted for easy mobilization. Figure 11-7 shows a typical system for an individually compressed low pressure well on gas lift. This might be a system on a gas well to help lift liquids.

11.7.2 Chamber Lift Installations

When the completion configuration prevents the point of injection from achieving the desired depth or when the volume of gas in an

Gas lift –
retrievable valves
chamber installation

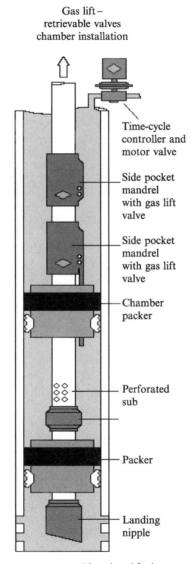

Time-cycle
controller and
motor valve

Side pocket
mandrel
with gas lift
valve

Side pocket
mandrel
with gas lift
valve

Chamber
packer

Perforated
sub

Packer

Landing
nipple

Figure 11-8. Chamber lift design.

intermittent lift installation is less than acceptable, a chamber lift design is used (Figure 11-8).

The concept of chamber lift is to create a large diameter volume (chamber) to collect liquids. The larger diameter of the chamber as opposed to the tubing allows higher volumes of liquid to accumulate while keeping the liquid column height to a minimum. Lower liquid column heights put less hydrostatic pressure on the formation. Increasing the diameter of the chamber can drastically reduce the hydrostatic head because the bottomhole pressure is reduced by the square of the chamber diameter. For example, increasing the chamber diameter from 2 3/8 to 3 inches will drop the hydrostatic pressure at the bottom of the hole by almost half for the same volume of liquid.

Typically, the chamber consists of a portion of the casing as shown in Figure 11-8. Chamber packers isolate the chamber, and a dip tube is frequently used in the top packer to allow the gas collected in the chamber to bleed off into the casing above the packer. Chambers can be manufactured at the surface and installed in the tubing string.

Chamber lift is one method of producing a relatively high volume of liquids in a low-pressure formation without loss of gas production because of excessive liquid head in the tubing.

In Figure 11-9, a "chamber" is formed between two packers. Well liquids are allowed to enter the space between the packers at low pressure. After the chamber is filled, gas is injected into the top of the chamber, displacing the liquids into and up the tubing. An additional gas lift effect is added to the liquids as they rise with gas injected from gas lift valves spaced higher in the tubing. A time-cycle controller is provided to control the cycles.

11.7.3 Horizontal Well Installations

Over the past decade, the number of horizontal wells has ballooned worldwide. Many of these wells are on gas lift either to increase oil production or, in gas wells, to more effectively produce the liquids.

Some operators have attempted to install gas lift in the horizontal section of the hole but have found this to be impractical for a variety of reasons.

- Gas lift operates by reducing the hydrostatic head on the formation. In a horizontal or near-horizontal section of the hole, there is very little vertical head. Placing gas lift valves in the horizontal lateral provides little benefit.
- In the horizontal section of the well, the two-phase (gas/liquid) flow tends to become stratified, allowing the gas to pass over the top of the fluid without pushing the fluid to the surface. This greatly reduces the efficiency of the gas lift.

Loading problems with horizontal wells

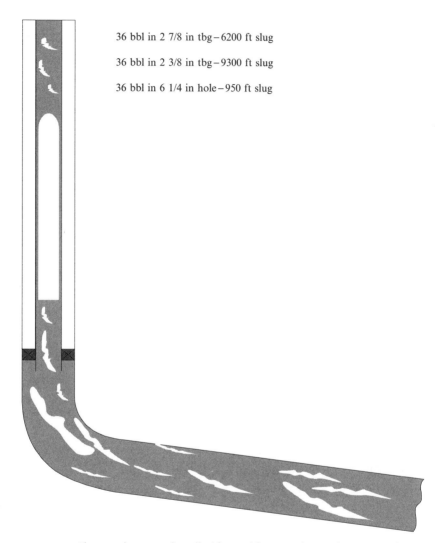

36 bbl in 2 7/8 in tbg – 6200 ft slug

36 bbl in 2 3/8 in tbg – 9300 ft slug

36 bbl in 6 1/4 in hole – 950 ft slug

Figure 11-9. Flowing horizontal well. (If gas lift is used to enhance production, install mandrels in the near-vertical portion only.)

- Servicing gas lift valves with slickline becomes increasingly difficult with increased wellbore inclination.
- Installing gas lift in the horizontal will, in general, not reduce the producing bottomhole pressure and could even increase it due to friction.

However, gas lift in the horizontal section may stabilize slugging and erratic flow.

The preferred completion configuration for a horizontal well on gas lift is to use the gas lift mandrels only in the portion of the wellbore where the deviation from vertical is less than 70 degrees. Figure 11-9 shows a horizontal well with inherent slugging in the vertical portion of the well. If gas lift is used, then gas lift valves should be placed in the vertical or near-vertical portion of the well and not in the horizontal section. This eliminates the problems discussed previously while producing the well at an optimum rate and efficiency.

Horizontal wells are also notorious for slugging, which dramatically reduces the overall production. Slugging can also create many problems with pumping systems, such as ESPs and beam pumps, where the slugging typically causes intermittent shutdowns in the equipment and cooling problems with the ESP motors. Slugging in the near-vertical portion of a horizontal flowing well is shown in Figure 11-9.

Installing gas lift in a horizontal well can stabilize slugging and thereby increase production. Gas lift removes the liquid head and controls the influx of gas to prevent or drastically reduce slugging, returning the production stream to more continuous flow. Gas lift also is not as affected by slugging as are most other mechanical pumping methods.

11.7.4 Coiled Tubing Gas Lift Completions

To reduce costs while improving the versatility of gas lift systems, coiled tubing suppliers have developed spoolable systems that can be run on coiled tubing. These systems provide complete downhole assemblies that can be installed in small-diameter casing or even tubing strings. The system can save initial costs with rigless completions and lower installations times. The cost of the coiled tubing gas lift string is comparable to a jointed tubing installation. The smaller diameter coiled tubing can also improve the efficiency of the lifting process by reducing the overall area of the pipe but at the expense of the added frictional drag imposed by the smaller cross-sectional flow area.

Coiled tubing gas lift completions have been available for nearly a decade. Some successful installations have been installed. Development of the systems continues to improve string reliability and better algorithms to predict the depth of the gas lift valves that include the sometimes significant stretch of the coiled tubing.

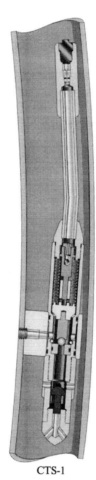

CTS-1

Figure 11-10. Spoolable coil tubing gas lift system. (Courtesy Schlumberger.)

Figure 11-10 shows a typical spoolable gas lift system with a close-up view of the spoolable gas lift valves. This system has the valves made up inside the CT during run-in, and the CT must be retrieved if the valves are to be serviced. Figure 11-11 shows a system by Nowcam (now Schlumberger) whereby the valves are inside the CT but the valves can be serviced by wireline.

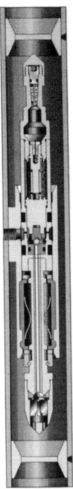

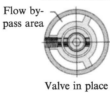

Flow by-
pass area

Valve in place

Figure 11-11. Valve in CT that is wireline retrievable. (Courtesy Schlumberger.)

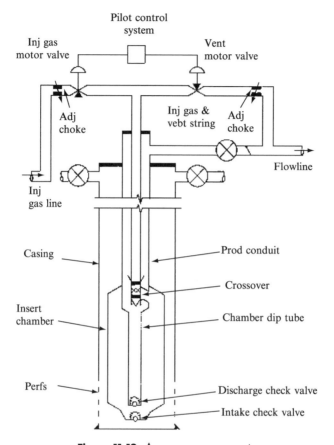

Figure 11-12. A gas pump concept.

11.7.5 Gas Pump Concept

A gas pump (Figure 11-12) is a form of intermittent gas lift where the injection gas does not mix with the produced liquids.[4] Although developed for viscous oil, this method can also be used for de-watering gas wells.

The gas pump is applicable only for shallow gas wells where there is sufficient injection pressure to overcome a hydrostatic gradient to the bottom of the well. The gas pump is a form of chamber lift in that a large downhole chamber is used to collect the liquids before being pushed to the surface by the gas. Although this method requires high-pressure gas at the surface, the volume of lift gas required is small compared to conventional intermittent gas lift systems.

Operation of the system begins with the chamber filling with produced liquids. After a predetermined time, high-pressure gas is injected rapidly into the chamber, forcing the liquid into the production tubing. During the injection process, the liquid is pushed into the production conduit with very little of the injection gas. An intake check valve closes during gas injection to prevent backflow into the formation. Once the injection gas begins to break around the bottom of the chamber, the well is shut-in and again allowed to accumulate liquids. The cycle is then repeated.

11.7.6 Gas Circulation

Another method of controlling liquid loading is to continuously inject gas down the casing and up the tubing to keep the gas velocity above the critical velocity at all times.[2] No gas lift valves are used. Figure 11-13 shows a schematic where this is done by compressing some of the gas back down the annulus.

Figure 11-14 shows how this is done with injection and how to use a compressor to reduce the wellhead pressure.

Figure 11-15 shows how a compressor can be used to lower the wellhead pressure and also inject gas downhole to stay above the critical velocity.

11.8 GAS LIFT WITHOUT GAS LIFT VALVES

It is possible to install holes in tubing with a check valve and a hard orifice[5] using wireline, as shown in Figure 11-16.

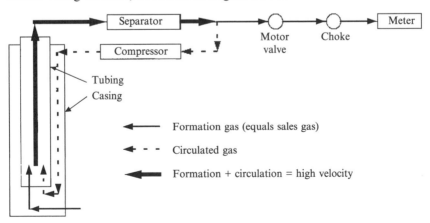

Figure 11-13. Gas injection to stay above critical velocity in tubing.

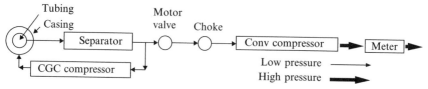

Figure 11–14. Gas injection to stay above critical velocity in tubing with gas compressor and wellhead compression.

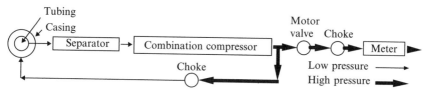

Figure 11-15. Gas injection to stay above critical velocity in tubing with single compressor.

This method allows gas to be injected into the tubing without having to pull the tubing and install mandrels and valves. It is "shot" though the tubing using a charge somewhat like shooting a perforation.

The gas cannot be injected as deep as would be possible using gas lift valves because there is no closing mechanism for the wireline set "seats." However, if one "seat" is installed, this will allow lightening the tubing gradient above this point when gas is injected, and then a second and possibly a third seat can be installed for deeper injection. However, this will result in multiple point injection as the upper "seats" cannot be closed. This is an inexpensive method of trying gas lift to see if you can unload a loaded gas well. Later, conventional gas lift could be installed if deeper injection is needed for improved production.

11.9 SUMMARY

- Gas lift for gas wells can be thought of as a method to keep the gas velocity above the critical velocity at all times. If this is done, then no liquid loading can occur.
- Intermittent methods use a burst of gas to lift liquid slugs from the well.
- Chambers allow liquid accumulations to occur for lifting with a minimum pressure on the formation.
- Liquids can be "pushed" to the surface using chamber type installations, but the gas pressure available must overcome a top-to-bottom hydrostatic liquid gradient.

Figure 11-16. A check valve and orifice for gas lift installed in tubing using wireline techniques.

REFERENCES

1. Trammel, P., and Praisnar, A., "Continuous Removal of Liquids from Gas Wells by Use of Gas Lift," SWPSC, Lubbock, TX, 1976, p. 139.

2. Boswell, J. T., and Hacksma, J. D., "Controlling Liquid Load-Up with Continuous Gas Circulation," SPE 37426, presented at the Production Operations Symposium, Oklahoma City, OK, March 9–11, 1997.

3. Stephenson, G. B., and Rouen, B., "Gas-Well Dewatering: A Coordinated Approach," SPE 58984, presented at the SPE International Petroleum Conference and Exhibition in Villahemosa, Mexico, February 1–3, 2000.

4. Winkler, H. W., "Gas Lift Solves Special Producing Problems,"*World Oil*, November 1998, 209, No. 11, pp. 35–39.

5. J. C. Kinley, Co., 5815 Royalton St., Houston, TX.

ELECTRIC SUBMERSIBLE PUMPS

12.1 INTRODUCTION

Electric submersible pumps (ESPs) are typically reserved for applications where the produced flow is primarily liquid. High volumes of gas inside an electrical pump can cause gas interference or severe damage if the ESP installation is not designed properly. Free gas dramatically reduces the head produced by an ESP and may prevent the pumped liquid from reaching the surface. In gas reservoirs that produce high volumes of liquids, ESP installations can be designed to effectively remove the liquids from the wells while allowing the gas to flow freely to the surface.

This chapter discusses the three main methods that use ESPs to de-water gas wells.

1. The first method develops techniques to separate the gas from the intake of the ESP so that liquid primarily enters the pump. The gas separation is accomplished by using completions or special separation devices. In this way, the liquid is produced to surface through the tubing, and the gas is allowed to flow freely up the annulus between the tubing and casing.
2. Another method is to use special stages at the pump intake to handle the gas. The special stages build pressure from the intake to compress the gas sufficiently so that conventional stages take over and can continue building pressure. This allows the ESP to pump with a fairly reasonable volume of free gas through the pump with the early special stages.
3. The third method is a technique where the liquid is re-injected into a formation below the packer. In this method, the liquid never reaches

the surface. If the pump is well below the gas perforations, the water falls by gravity to the pump intake while the gas flows up the annulus. This system is commercially available and has been used in several successful gas well de-watering installations.

12.2 ESP SYSTEM

This chapter is not intended as a tutorial on the installation of ESP systems. Some basic knowledge of ESPs is assumed. Some introductory comments are included describing only the basics of an ESP installation. See Reference 3 for more details.

Figure 12-1 shows an approximate depth-rate chart for ESP applications. This chart suggests that ESP applications can be made inside the envelope and not outside of the envelope. However, there are many exceptions to the chart possibilities that would extend or reduce the area indicated as possible for applications.

Figure 12-2 shows a basic ESP system. The system consists of a down-hole motor connected to a seal section, which in turn is attached to a pump intake and then a centrifugal pump. A high voltage three-phase electric cable connects the motor to the surface where either a high voltage transformer or a variable speed drive (VSD) transformer supplies the electrical power.

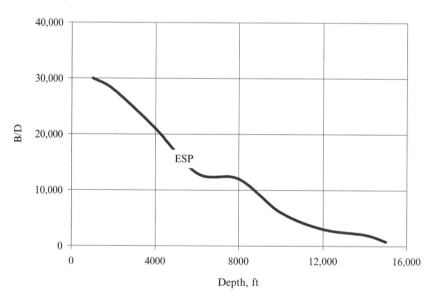

Figure 12-1. An approximate depth-rate application chart for ESPs.

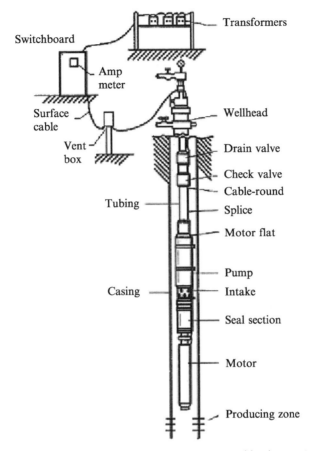

Figure 12-2. Typical ESP system. (Courtesy Schlumberger.)

The motor is a two-pole, squirrel cage motor with a synchronous speed of 3600 RPM and an operational speed of approximately 3500 RPM at 60 Hz. It is imperative that the motor be cooled by the produced fluid passing its outer casing. In the event that large quantities of gas pass the motor, the heat transfer from the motor to the produced fluid will be drastically reduced, potentially causing severe motor damage.

The seal section houses a pump thrust bearing and restricts the wellbore fluids from entering the motor. The pump has an intake where the fluid enters the pump at the bottom of the pump. The intake can be replaced by a rotary gas separator, which separates gas to the annulus while nearly all liquid enters the pump. The pump itself consists of a stack of impeller/diffuser combinations that generate head. The amount of head required to bring the liquids to the surface dictates the numbers of impeller/diffuser pairs, and the flow rate required determines what type (size for a particular narrowflow rate range) of stages to use.

The motor controller typically has protective shut-offs, the on/off controls, usually some method of recording motor amps, and other parameters often used to supply diagnostics for the pump operation. The transformers drop the voltage from the transmission line voltage to the value needed by the motor after accounting for the cable voltage loss.

A typical pump performance curve for a single pump stage is shown in Figure 12-3. This curve shows that the pump reaches maximum efficiency at a rate of approximately 1200 bpd and produces about 26 ft of head per stage. Higher head is possible at lower flow rates, and more flow is possible at significantly lower values of head.

The single stage head curve shown below is for the special case where the pump is pumping 100% liquid. If gas is present, the head curve tends to become erratic and may drop to zero head prematurely as the flow rate of gas is increased. It is therefore necessary to keep large quantities of gas from entering the pump intake of an ESP system.

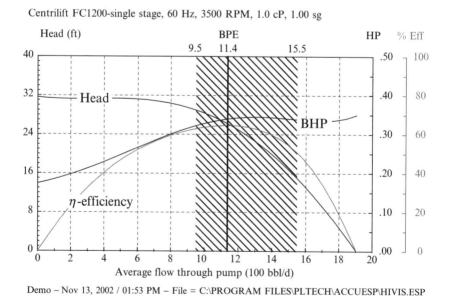

Centrilift FC1200-single stage, 60 Hz, 3500 RPM, 1.0 cP, 1.00 sg

Demo – Nov 13, 2002 / 01:53 PM – File = C:\PROGRAM FILES\PLTECH\ACCUESP\HIVIS.ESP

Figure 12-3. Pump performance curve showing the head curve, the brake HP, or BHP curve, and the stage efficiency curve.

12.3 WHAT IS A "GASSY" WELL?

ESP performance can be severely degraded by the flow of excessive gas through the pump. But what constitutes excessive gas? How much free gas can a given ESP handle before performance is affected?

Although a high GOR or GLR could be an indicator of excessive gas, a high intake pressure would compress the free gas, making the free gas volume smaller, whereas a low intake pressure would expand the free gas resulting in higher gas volume through the pump. In addition, if the intake pressure is greater than the bubble point, all the gas will be in solution so that no free gas will be present in the pump.

This section will summarize a method for evaluating the effects of free gas on ESP performance. See Reference 1 for more details.

A useful approximate correlation for evaluating ESP performance with free gas is[2]:

$$\Phi = \frac{666 \times \text{VLR}}{P_{\text{ip}}}$$

where VLR = vapor/liquid volume ratio at the pump intake
P_{ip} = pump intake pressure, psia

It is found empirically that the effect of free gas on the pump head curve is negligible when Φ is approximately less than 1.

Example 12.1: Calculate Free Gas Percentage and ESP Limitations:

Desired production	1000 STB/D
Pump intake pressure	850 psia
Pump intake temperature	165°F
Produced GOR	430 scf/bbl
Gas gravity	0.65
Water gravity	1.08
Oil gravity	35 API
Water cut	65%

In this example, we will use Standing's Black Oil correlation[3] for solution gas R_s and oil formation volume factor B_o.

$$R_S = \gamma_G \left(\frac{P_{ip} 10^{0.0125API}}{18 \times 10^{0.00091T}} \right)^{1.2048} \text{ scf/STB} - \text{oil}$$

$$F = R_S \left(\frac{\gamma_g}{\gamma_0} \right)^{0.5} + 1.25T$$

$$Bo = 0.972 + .000147F^{1.175} \text{ bbl/STB}$$

1. First calculate the gas in solution at the pump intake from Standing's[3] solution GOR, scf/bbl-oil:

$$R_S = .65 \left(\frac{850 \times 10^{.0125 \times 35}}{18 \times 10^{.00091 \times 165}} \right)^{1.2048} = 150 \text{ scf/STB} - \text{oil}$$

2. Calculate the formation volume factor B_0 following Standing's[3]

3. $\gamma_0 = 141.5/(131.5 + 35) = 0.85$

$$F = 150 \left(\frac{.65}{.85} \right)^{0.5} + 1.25 \times 165 = 337.43$$

$$B_0 = 0.972 + .000147F^{1.175} = 1.11 \text{ bbl/STB}$$

4. Calculate the gas volume factor at the pump intake using $Z = 0.85$ for the compressibility factor:

$$B_G = \frac{5.04ZT(^\circ R)}{P_{ip}} = \frac{5.04 \times .85 \times (165 + 460)}{850} = 3.15 \text{ bbls gas/Mscf}$$

5. Calculate the volume of free gas in in-situ barrels and the oil and water volumes at the pump intake. The total free gas at the pump is then:

$$Q_{gas} = Q_{oil} \left(\frac{GOR - R_S}{1000} \right) B_g = 350 \times \left(\frac{430 - 150}{1000} \right) \times 3.15$$
$$= 308.7 \text{ bbls gas/day}$$

Assume 30% of the free gas at the pump travels up the annulus, bypassing the pump intake. Then only 70% of the free gas actually enters the pump:

$$Q_{gas,pump} = 0.7 \times 308.7 = 216 \text{ bbls gas/day}$$

6. Calculate the volume of liquids entering the pump.

$$Q_{oil,pump} = Q_{oil,STB} \times B_0 = 350 \times 1.11 = 388.5 \text{ bbls oil/day}$$

$$Q_{water,pump} = 650 \text{ bbls water/day}$$

7. Calculate the parameter Φ

$$\Phi = \frac{666(Q_{gas}/Q_{liquid})}{P_{ip}} = \frac{666}{850} \times \frac{216}{388.5 + 650} = 0.164$$

If $\Phi < 1.0$, then the ESP will perform on the nominal head curve, even though some free gas is present.

If $\Phi > 1.0$, then the pump head curve will be degraded, and a gas separator will be needed to reduce free gas through the pump. The gas separator would augment the natural separation of the free gas bypassing the pump into the annulus.

In this example, $F < 1.0$, so no additional gas separation is predicted.

8. Calculate the percent free gas at the pump.

$$\% \text{ Free Gas} = \frac{Q_{gas}}{Q_{gas} + Q_{liquid}} \times 100 = \frac{216}{216 + 388.5 + 650} 100 = 17\%$$

The percentage of free gas is fairly high but the high pump intake pressure of 850 psia reduces the detrimental effects of free gas on the pump performance.

12.4 COMPLETIONS AND SEPARATORS

As we have seen, excessive gas at the pump may require additional gas separation to reduce the free gas into the pump. This section will discuss methods to achieve higher gas separation. Perhaps the best method of keeping gas from entering the pump is to set the pump intake below the perforations. This configuration would allow the liquids to gravity drain to the pump intake while the lighter gas is diverted into the annulus above the pump. However, this completion locates the ESP motor outside the flow path of the production liquids, where it would not normally receive sufficient cooling. To alleviate this problem, the motor can be fitted with a shroud that forces the produced liquids, down past the motor before entering the pump intake as illustrated in Figure 12-4.

If the pump must be set above the perforations, then the pump can be fitted with an upward opening shroud on the pump intake, also shown in Figure 12-4. This forces the produced fluid to reverse direction and travel downward before entering the pump, breaking out the larger gas bubbles. In high rate wells, however, where the downward velocity of the liquid in the annulus between the casing and the shroud is above approximately 1/2 ft/sec, the production fluids can carry significant amounts of free gas to the pump intake even with shrouds.

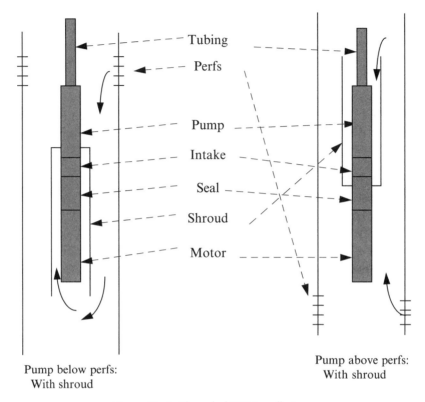

Tubing

Perfs

Pump

Intake

Seal

Shroud

Motor

Pump below perfs:
With shroud

Pump above perfs:
With shroud

Figure 12-4. Shrouded ESP installations.

Although shrouds can increase gas separation, several potential problems should be considered before running shrouds on ESPs.

- A shroud can substantially decrease the clearance between the pump assembly and the casing. In wells having clearance problems, a full-length gauge section should be run before running the pump.
- The shroud can accumulate sand, scale, and asphaltenes. This is particularly true for upward facing shrouds, which act like collectors for heavy particles.

If the annular region between the shroud and the motor (or pump) is small, the increased pressure drop inside the shroud can reduce the pump intake pressure and break gas out of the solution just ahead of the pump intake. Other methods exist, such as devices to re-circulate fluids to the motor, when the motor is set below the perforations.

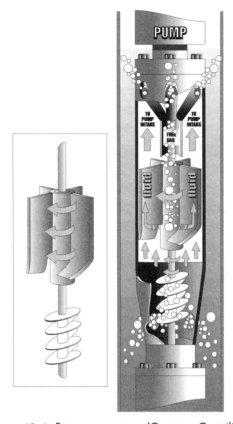

Figure 12-5. Rotary separator. (Courtesy Centrilift.)

Another common device used to remove gas from the production stream before entering the pump is the "rotary separator" (Figure 12-5). This device is fitted to the pump intake and is attached to the rotating pump shaft. The centrifugal action of the separator causes gas to be diverted to the annulus, leaving mainly liquid to enter the pump. Tests have shown that the rotary separator can be more than 90% effective. The rotary separator, can also be gas choked, however, if gas volumes become too large and rates too high. It may wear with sand because there are large velocity gradients where the fluids are swirled. See References 4, 5, and 6 for more information on various gas separation and handling devices applied in the industry.

Some pump suppliers have stages that are designed to better handle free gas. These stages are installed at the pump intake and build pressure in the presence of free gas. After the gas is sufficiently compressed, additional standard pump stages are used to generate the desired pump discharge pressure.

Examples of these special "gas handling" stages are the Schlumberger Advanced Gas Handler[7] (AGH) and the Centrilift Gas Master[4] stage. The AGH has a recirculation path to keep bubbles from accumulating and to keep some pressure building in the pump. A few of these stages in a pump in a gassy environment, even pumping under a packer, may be able to build pressure in the pump so that gas is not a problem for the remainder of conventional stages in the pump assembly.

A simpler option is to use the "tapered pump" concept of design with larger stages at the intake to accommodate free gas and switching to smaller stages as compression reduces the total volume that can help to handle gas. This option may require many stages compared to fewer special stages.

12.5 INJECTION OF PRODUCED WATER

ESPs can also be used to re-inject liquids back into the formation in a manner similar to that performed by beam pumps (see Chapter 10). In this system (illustrated in Figure 12-6), the ESP is inverted to push fluids downward. The pump generates the necessary head to push the liquids into a water injection zone below the packer while the gas is produced up the casing annulus. No liquids are produced to the surface. The system is usually installed with a downhole pressure sensor that detects when a predetermined level of liquids has accumulated over the pump. Once this level is reached, the sensor starts the ESP motor to inject the liquids; once the liquids have been injected, the liquid level drops and the sensor automatically shuts off the ESP.

Be aware that many injection systems are operating above the "parting pressure" of the formation. Depending on formation characteristics, a continually lengthening fracture may result, which may later intersect a well or producing interval with undesirable effects.

12.6 SUMMARY

- ESPs can be a viable method to de-water gas wells, usually when it is necessary to handle large liquid volumes. Generally, ESPs are considered only when water rates exceed at least 100 bpd. But if high rates are needed, they become much more advantageous. Small water well ESPs are used to lift relatively small rates off of coal gas wells.

Centrilift gas prosystem

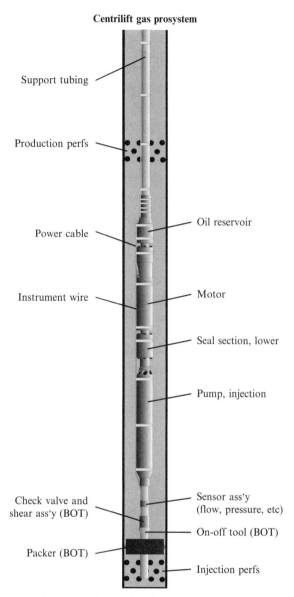

Support tubing

Production perfs

Power cable — Oil reservoir

Instrument wire — Motor

Seal section, lower

Pump, injection

Check valve and shear ass'y (BOT) — Sensor ass'y (flow, pressure, etc)

On-off tool (BOT)

Packer (BOT) — Injection perfs

Figure 12-6. Inverted ESP installation for de-watering gas wells by injecting water below a packer into a water zone.

- ESP installations are expensive and usually consume a little more power per barrel of liquid lifted than a beam pump system. Of course, they should be compared only when the rates are well within the good operational ranges for both the beam and ESP systems.
- The efficiency of an ESP system is significantly reduced (similarly for a beam system and other systems, excluding gas lift) when gas is allowed to enter the pump. These shortcomings limit the use of ESPs for gas well de-watering applications.
- The use of ESPs to inject water below a packer at fairly high rates is a specialty area for ESPs for gas well operations.

REFERENCES

1. Centrilift Submersible Pump Handbook, 6th Edition, Centrilift, Claremore, OK, 1997.

2. Turpin, J. L., Lea, J. F., and Bearden, J. L., "Gas Liquid Flow through Centrifugal Pumps-Correlation of Data," Proc. 33rd Annual Meeting of SWPSC, Lubbock, TX, 1986.

3. Standing, M.B., "Volumetric and Phase Behaviors of Oil Field Hydrocarbon Systems," Reinhold, New York, 1952.

4. Centrilift Gas Handling Manual.

5. Dunbar, C.E., "Determination of Proper Type of Gas Separator," REDA Technical Bulletin, 1989.

6. Wood Group ESP Inc. Bulletin on XGC gas separator, Wood Group ESP, Oklahoma City, OK, 2002.

7. Schlumerger-REDA bulletin on the Advanced Gas Handler (AGH) stages, Schlumberger-REDA, Bartlesville, OK, 2001.

PROGRESSIVE CAVITY PUMPS

13.1 INTRODUCTION

Introduced in 1936, the progressive cavity pump (PCP) is of simple design and has the ability to handle solids and viscous fluids required for many applications.

According to ABB,[1] there are approximately 60,000 PCP applications in the world—with the most being in Canada and Venezuela. Many are used for de-watering coal bed methane production, primarily due to the solids handling ability, and other gas wells, as well for water and oil well production. For general de-watering of gas wells, PCPs might be recommended if:

- Depth is shallow (<4000 to 6000 ft)
- Fluid rates are relatively high
- Low profile pumping system is required
- Good power efficiency is sought
- Solids are in the production
- Well temperature is low

Figure 13-1 shows approximate depth-rate limitations for the PCP system.

Figure 13-1 illustrates the design of the PCP. The pump has only one moving part downhole and no valves. The pump will not gas-lock, but the elastomeric stator can overheat handling gas. It can produce sandy and abrasive formation fluids and is not usually plugged by solids.

A PCP does have limitations. The rubber or elastomeric stator may be susceptible to chemical attack and high temperature and is generally

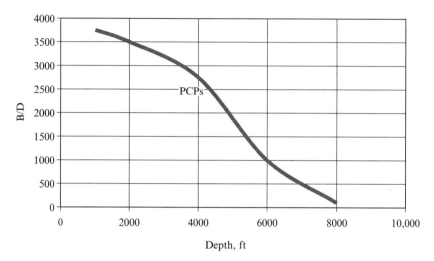

Figure 13-1. An approximate depth-rate envelope for PCP system applications (Courtesy Weatherford.)

limited to depths less than approximately 4000 to 6000 ft. Catalogs show PCP performance to ≈ 6000 to 7000 ft, and although possible, operational problems may be severe. A cross-section of the pump is shown in Figure 13-2, illustrating the cavities carrying fluid while pumping. The pumping rate can be easily adjusted to well conditions by changing the pump RPM.

The PCP unit consists of two main parts: a moving single helical steel rotor and a stationary double-threaded helical elastomer stator. With the rotor in the stator, a series of sealed cavities are formed. As the rotor turns, the cavities progress in an upward direction (Figure 13-3). A cross-section of the rotor inside the stator is shown in Figure 13-4.

Two variables should be considered when matching the pump to well conditions. The first is pump capacity, which is determined by the size of the cavities formed between the rotor and stator. Larger cavities produce higher flow rates at a given well depth and rate of rotation. The second is depth capability, which is determined by the number of seal lines controlled by the length of the rotor and stator. A longer rotor and stator with a shorter pitch will allow a PCP pump to pump from greater depths at higher given capacity rating.

Commercially available PCP pumps can be found to range in performance from 3000 bpd at 2000 ft depth down to 600 bpd or less at 6000 feet. Longer life and better efficiency are usually found with shallower installations.

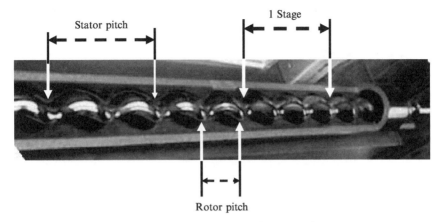

Figure 13-2. Progressive cavity rotor/stator pitch geometry.

Figure 13-5 shows a PCP installation with an electrical motor and sheaves and belts to perform the needed speed reduction. An alternative to belt drives is a hydraulic drive (not shown), which will allow continuously variable speeds to maintain a proper fluid level over the pump.

13.2 PCP SYSTEM SELECTION

The following briefly outlines some considerations for each component of a PCP system, primarily from Reference 2.

13.2.1 Rotor

The rotor is normally made from alloy steel equivalent to 4140 or 4150 carbon steel, with a Brinell hardness of 200 to 240 and a heavy layer of chrome plate. Alternate coatings include tungsten carbide, boron, nickel, and ceramic; 300 and 400 series stainless steel and 17–4 pH stainless have been applied in very corrosive environments. Well liquids and solids produced should be analyzed when selecting rotor materials.

Chrome plating can be used if the pH of well fluids is between 5 and 8. If the pH is below 5, then stainless steel should be selected to minimize corrosion. Most stainless steels are soft compared to 4140 alloy steel; therefore, the percentage of solids and the abrasive nature of the solid particles must also be considered before specifying the type of stainless steel. Use of 316 stainless is most common; however, it is the softest of the other materials, such as 416 or 17–4 pH. Materials, such as a

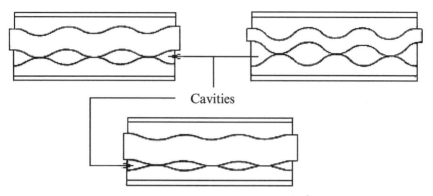

Figure 13-3. PCP cavity pump cycle.[2]

boronized rotor, are both corrosion and abrasion resistant and can yield run lives of three to five times that of chrome in the same application but are more costly.

Rotors having thin chrome coatings (0.010 inch) will sustain damage to the base metal of the rotor before the loss of efficiency is detected and therefore cannot often be re-plated. Thicker platings (0.016 to 0.020 inch per side) are more effective as the loss in pump performance is detectable before base metal rotor damage occurs.

Examples (but not comprehensive) of industry options include:
Manufacturer B Models

- 350–4100 HN Stator and Chrome Rotor
- 400–4100 HN Stator and Chrome Rotor

Manufacturer G Models

- 340–4000 B Stator and Chrome Rotor
- 400–4250 B Stator and Chrome Rotor

Manufacturer GE Models

- 20.40–2100 NBRA Stator and Chrome Rotor

13.2.2 Stator

The stator is molded using an elastomer compound. Most are nitrile based using various acrylonitrile (ACN) concentrations, additives, and cures that alter performance characteristics. Common compounds

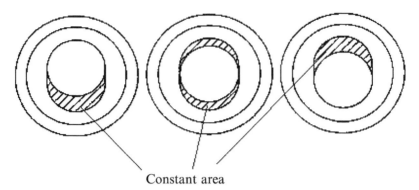

Constant area

Figure 13-4. Progressive cavity pump rotor/stator relationship.[2]

consist of low or medium high acrylonitriles (Buna N) high nitriles (increased ACN content), highly saturated hydrogenated nitriles (HSN), and some flourel formulations (vitons). Limitations include oil gravity (<40 API), temperature (<350°F), CO_2 concentration (<2% in-solution), H_2S concentration (<15% in-solution), and compatibility with treating and some EOR chemicals. In general, more ACN provides aromatics resistance but has less desirable mechanical properties.

Recommended immersion tests of elastomer samples in laboratory tests do not always indicate what difficulties may be encountered in the actual application.

Nitrile-based compounds are the standard, with the main difference being the ACN content.

Buna N or low to medium acrylonitrile (ACN Nitriles)

With typical oils having less than 25 API, Buna N or low to medium high acrylonitrile can be used. It performs well in high water cut applications, such as de-watering gas wells, and also with CO_2. The temperature limitation in wells is approximately 180°F. Do not use with H_2S.

High (ACN) nitriles

This type of elastomer performs similarly to Buna. Increased ACN content gives aromatic swell resistance yields and better performance in oils with API >25 and up to approximately 38 API. Check for

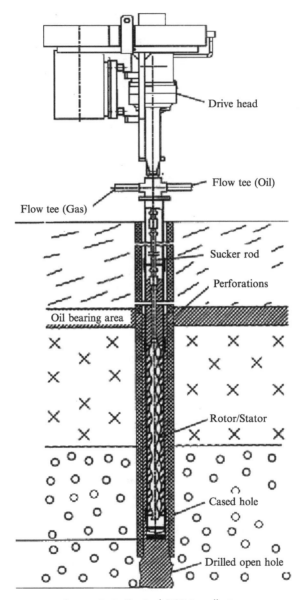

Figure 13-5. Typical PCP installation.

possible swelling in well fluids. It is not recommended for H_2S service. The temperature limitation is approximately 200 °F.

Highly saturated, hydrogenated nitriles (HSN)

This elastomer is used mainly with H_2S and higher temperature. H_2S resistance varies with the compounds but has been used to 15% in-solution. The temperature limitation is compound dependent and ranges from 250 °F to 300 °F. Aromatic resistance is less than for high nitrile.

13.2.3 Surface Drive

The surface drive supports the weight of the rod string and fluid and supplies power to the pump through the rod string. The bearing housing of the gear head supports the weight of the rods and the hydraulic load. The drive shaft can be turned using belts and sheaves or by connecting the prime mover to the drive shaft. The prime mover can be an electric motor, hydraulic motor, gas, gasoline, or diesel engine. The gear head can be either solid or hollow shaft. The hollow shaft allows lifting the rotor out of the stator without removing the gear head from the wellhead.

13.3 SELECTION AND OPERATIONAL FACTORS[1]

13.3.1 Important Factors for Sizing the System

The following section highlights some of the important considerations for sizing a PCP installation.

- The two basic conditions required to size the downhole pump are total dynamic head (TDH) and desired flow rate. The TDH is the pressure drop across the pump expressed as an equivalent vertical height of the fluid being pumped.

$$TDH = \frac{\Delta P_{\text{Pump}}}{0.433 \ \gamma_f} = \frac{P_D - P_{\text{pip}}}{0.433 \ \gamma_f} \ ft$$

where P_D = pump discharge pressure, psi
P_{pip} = pump intake pressure, psi
γ_f = fluid specific gravity

1. The P_D includes the weight of the fluid in the tubing, surface tubing pressure, and friction in the tubing. The P_{pip} is a result of the inflow of the well and may be measured by including the surface casing pressure, the weight of gas in the casing/tubing annulus, and the corrected effects of a fluid level over the pump. The gas flowing up the annulus should always be free to pass through a check valve and into the tubing at the surface for any pumping system.
2. Determine the abrasive characteristics of the fluid. If the solids concentration is considered heavy, avoid running the pump in the upper half of the possible pump speed range.
3. Choose a pump model that fits the requirements for pressure/stage, speed/abrasion, and desired flow rate. One method of sizing is to choose a pump with a fit that will give approximately 40% to 50% efficiency in the shop. Then, as the elastomer swells in the field application, aim for 70% or more as measured in the field. The pump efficiency is defined by:

$$\eta = \frac{Q}{K_{Pump} \times RPM_{PR}} \times 100$$

where
$$Q = \text{actual pump rate, bbl/day}$$
$$K_{Pump} = \text{pump design constant, bbl/day/RPM}$$
$$RPM_{PR} = \text{RPM at polish rod}$$

Example 13.1: PCP Efficiency

A pump has a design constant of 1.0 bpd/RPM. The reported production of a high water cut well is 186 bpd. It is desired to maintain about 70% efficiency in field operations. Is this being accomplished? The PCP is turning at 200 RPM.
Solution:
The efficiency can be calculated as:

$$\eta = \frac{186}{1.0 \times 200} \times 100 = 93\%$$

This is well above the required 70% for good field efficiency. It may be too good if the torque, and especially the starting torque, is rising above acceptable limits. See previous recommendations to start with a low efficiency, and let stator swelling bring the pump into a good efficiency range as time progresses.

Different-sized rotors are used to get the fit at a given temperature (i.e., B, C, D, E, F, G, or H). Each rotor fit is about .004 inch less than the previous. Experience and databases show which rotor to select for each application.

The sample performance curves in Figure 13-6 show how the production drops off with increase in depth due to the increased slippage of fluid through the rotor/stator.

13.3.2 Steps to Size the PCP

1. Check to see that the selected pump will fit in the casing.
2. Examine the performance curve for a pump of the required fit to determine the horsepower and speed required for the application. Use the starting torque to determine prime mover size required. If variable speed is utilized, the prime mover can be selected using the running torque. If a fixed speed system is used, the prime mover should be sized using the starting torque.
3. Determine the rod string size by using published data or calculate the maximum shear stress (S_S) of the rod due to combined axial and torsional loads using the equation

$$S_S = (A + .5B)^{.5} \text{ psi}$$

$$A = \left(\frac{16T}{\pi d^3}\right)^2$$

$$B = \left(\frac{(1. - .128\gamma_f) \ W_R L + A_P(.433 D \gamma_f + P_{Surf})}{A_{ROD}}\right)^2$$

where T = torque, in/lbf
$\quad\quad d$ = rod diameter, in
$\quad\quad W_R$ = rod weight in air, lbf/ft
$\quad\quad L$ = rod string length, ft
$\quad\quad D$ = pump vertical depth–fluid level over pump, ft
$\quad\quad \gamma_f$ = fluid specific gravity

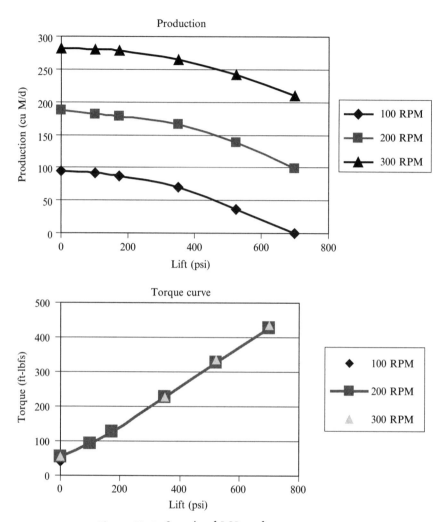

Figure 13-6. Sample of PCP performance curves.

P_{Surf} = surface tubing pressure, psi
A_{ROD} = rod cross-sectional area, in^2
A_{P} = effective rotor area, in^2
  = crest-to-crest diameter area minus the rod cross-sectional
    diameter area

The field proven limit for S_S is 30,000 psi for Grade D rod or equivalent.

Example 13.2: Rod Torque Limits

A PCP installation uses Grade D drive rods with a stress limit of 30,000 psi and has the following pump information.

T, torque	650 ft/lbf = 7800 in/lbs
d, rod diameter	.875 in
W_R, rod weight	2.22 lbf/ft
L, rod string length	2500 ft,
D, depth–fluid level	2000 ft
γ_f, fluid specific gravity	1.0
P_{Surf}, surface pressure	100 psi
A_{ROD}, rod area	.601 in^2
A_P, effective rotor area	.899 in^2

Are the rods overloaded?

Solution:

Calculate the maximum shear stress. If S_S is greater than the rod limit of 30,000 psi, it is considered an overload situation.

$$A = \left(\frac{16T}{\pi d^3}\right)^2 = \left(\frac{16 \times 7800}{3.1416 \times 875^3}\right)^2 = 3.516 \times 10^9$$

$$B = \left(\frac{.88 W_R L + A_p(.433 D \gamma_f + P_{Surf})}{A_{ROD}}\right)^2$$
$$= \left(\frac{.88 \times 2.22 \times 2500 + .899(.433 \times 2000 \times 1 + 100)}{.601}\right)^2$$
$$= 3.147 \times 10^7$$
$$S_S = (A + .5B)^{.5} = 59,428 \text{ psi}$$

The rod is overloaded as the maximum stress S_S is greater than the 30,000 psi limit. Try the calculation with a larger rod.

The data in this example are "running" values and without a variable speed drive. The starting torque can be approximated from the running torque by multiplying by 1.3, or 30%, more than the running torque.

The limiting shear stress is proportional to the yield strength of the rod, so compare the yield strength of the rod to the maximum shear limitation when evaluating other materials. Rod size is chosen by calculating the maximum shear stress <30,000 psi. If overloaded, choose the next larger rod diameter or a grade with a higher yield stress.

The rod selection graph[5] in Figure 13-7 provides a quick rod design. This chart shows an allowable torque of about 625 ft/lbs for Grade D at 2500 feet. Reference 6 has additional detail on selection of rods for PCP applications. The letters C, D, and K refer to common rod grades, and the numbers on the curves are the ODs of the various rod sizes.

4. Select a drive system and gearhead. Select a gearhead that has the thrust capacity that fits the application and will yield sufficient bearing life. Calculate the L10 life of the bearing to determine if the expected life is sufficient. Use the following equations to calculate the expected L10 life in hours:
   Calculate the axial load F_A that the bearing will need to support.

$$F_A = W_R(1 - .128\gamma_f)L + A_{ROD}(.433D\gamma_f + P_{SURF})$$

Then calculate the L10 life of the bearing in hours under these conditions.

$$Expected\ Life = 300\frac{L10}{F_A}\frac{500}{N}\ hrs$$

where $L10$ = axial bearing load rating, from manufacturer (lab)
     N = pump speed (RPM)

5. Select the drive system configuration. As mentioned earlier, this can be any type of prime mover. Be sure that the power source is sized correctly to start the unit.

Allowable red torques

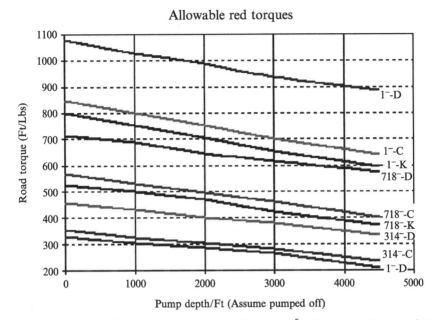

Figure 13-7. Allowable rod torques (after Griffin Pumps[5]). The letters C, D, and K refer to common rod grades. The numbers refer to the rod outside diameters.

6. Determine what ancillary equipment will be necessary to enhance the operation of the system by protecting the pump, tubing, rods, and maintenance personnel. The following section discusses pump-off controls, rod guides, special rod couplings, and different anchors used in the system.

13.4 ANCILLARY EQUIPMENT

Most PCP installations use additional equipment to enhance the efficient production of the liquids from the well. Many of these are necessary for normal operation because of the gassy conditions existing downhole. The following are a few of the most commonly found when PCPs are used to remove liquids from gas wells:

13.4.1 Flow Detection Devices

Flow detection devices should be considered if the design fluid level is expected to be near the intake of the pump or when the inflow of the well is not known, as in a new well.

This section briefly describes three common flow detectors. These devices can be instrumented with recorders and timers to control the pump operation. The systems can also be designed for manual restart or set to automatically restart after the well is given time to recover. There are other combinations and more sophisticated controllers and devices than those described here. Some operators simply shoot the fluid level periodically and adjust speed accordingly. Typically, devices for pump-off control should continue to show improvement in the industry.

Flow meters

Flow meters are used to measure the liquid production at the surface. Monitoring the fluid production permits the detection of changes in surface flow rate. Typically, the meter is set to trigger shut down when predetermined decrease in the flow rate occurs.

Differential pressure switches

A differential pressure switch is a configuration in the flowline where fluid flow is directed through an orifice. A given flow rate will result in a specific pressure drop across the orifice. This pressure drop or differential can be monitored through a pressure gauge, which has a preset low- and high-pressure shut off. As the flow rate decreases, the differential pressure across the orifice drops accordingly, and once the low set point is met the unit is shut down.

Thermal dispersion devices

Thermal dispersion devices are also flowline-mounted systems. The sensor, consisting of two thermal probes, is mounted in the flowline, such that a full stream of fluid passes by both probes. The temperature variance between the two probes is dependent of the nature of the fluid and the flow rate between the probes. The temperature correlates to a measured milliamp signal that can be recorded. A low milliamp signal then indicates a decrease in flow or a change in the medium passing through (i.e., a change from liquid to gas). When the preset low milliamp signal is reached, the unit can be shut down.

13.4.2 Rod Guides

Although rod guides are used primarily in deviated well applications, many producers routinely use them as a precaution against tubing wear. The number and location of the guides is determined by the severity of the deviation.

Commonly used are rod centralizers that fit in the tubing while the rod spins inside the centralizer. These both decrease the drag inside the tubing and eliminate any problems of tubing/rod guide wear.

Coated rod couplings also may be used. The coupling surfaces are either plastic, hard metal–faced, or rubber. These couplings are used in conjunction with the rod guides on the body of the rod for added protection. Monitor rod wear locations when pulling equipment to see where protection is needed.

13.4.3 Gas Separators

Gas separators should be considered in any application where the gas may be produced through the pump. The amount of gas passing through the pump has a direct effect on the pump volumetric efficiency. Several gas separator designs are currently available; each has specific advantages and disadvantages. The individual separator is the primary criterion used in its selection. Although the conventional "poor-boy" style separator is one of the most common in the field (see Chapter 10), it is among the least efficient.

If the pump is allowed to produce gas, adiabatic compression of the gas can quickly generate enough heat to damage the rotor elastomer. Friction plays some factor as lubrication is reduced. A fluid level must be maintained. If the stator-rator interface becomes too dry, friction also adds to heating.

Most of the discharge temperatures predicted[2] are harmful to the PCP stator elastomer. Adiabatic gas compression of a hydrocarbon gas is used to calculate the temperatures indicated in the table. If no liquids are being pumped, the high gas compression temperatures will damage or melt the stator in a short time.

The best separation technique is to set the pump intake as far below perforations as possible. This allows for the gas to break out of solution and migrate into the casing annulus before reaching the pump intake.

The equation used for the above calculations is:

$$T_{out} = T_{in} \left(\frac{P_{out}}{P_{in}} \right)^{\frac{k-1}{k}}$$

where k (≈ 1.3) is the ratio of specific heats for hydrocarbon gas.

13.4.4 Tubing Anchor/Catcher

Tubing anchors are used to ensure that the tubing will not come unscrewed during operation. Most applications do not require anchoring the tubing because it is unlikely that the prime mover has enough torque capability to unscrew the tubing if the tubing threads are in good condition and tightened to at least API specs. A tubing anchor will not serve as a tubing catcher, and catchers may be preferred as insurance against accidents happening. A tubing catcher can serve as a tubing anchor. Variable speed drive systems do offer the advantage of very high torque at low speeds. When using variable speed drives tubing anchors may be necessary.

13.5 TROUBLESHOOTING PCP SYSTEMS[5]

This section describes some common difficulties encountered using PCPs and some of the more common approaches to solving those problems.

Problem: No production with the correct RPM

- Possible broken sucker rod—lift and check the rod string weight
- Rotor may be broken, especially if welded joints are in play
- Hole in the tubing—pressure up tubing and check casing pressure
- Tubing unscrewed or parted
- Rotor not engaged in the stator
- Stator was mistakenly run upside down or rotor was run through stator
- PCP may be worn out—excessive rotor wear caused by sand, coal fines, or other solids

- Stator elastomer may have deteriorated because of chemical attack, overpressure, or high temperature, possibly from excessive gas through the pump

Problem: No production with slower than required polish rod speed

- Rotor may be jammed against the tag bar (stop pin)—recheck rotor landing
- Elastomer is torn from engaging rotor too quickly
- Prime mover providing less than needed power due to under sizing or damage

Problem: No production—polish rod is turning; prime mover speed is low

- Prime mover is undersized
- Prime mover has become damaged

Problem: Lower than expected production with the correct polish rod speed

- Well PI overestimated—check fluid level
- Inflow restrictions at the perforations
- Pump intake is plugged
- Higher than expected GLR—reduced volumetric efficiency in the pump
- Flow restriction in tubing caused by oversized rod couplings or rod centralizers
- Rotor is not fully engaged in the stator
- Worn pump
- Excessive elastomer swell—torque should be indicator
- Hole in tubing

Problem: Low production with slower than expected polish rod speed

- Incorrect sheave sizing
- High rod torque—rotor/stator fit is being affected
- High sand or coal fines production
- Chemical/aromatics attack of stator elastomer
- Fuses/overloads incorrectly sized

- Low line voltage

Problem: Slugging production with low polish rod speed

- Rotor is running on top of the tag bar (stop pin)
- Intermittent solids in production is plugging the pump and tubing
- Stator elastomer swelling
- Prime mover is overloaded

Problem: Slugging production with the correct polish rod speed

- High GOR—increase submergence, monitor fluid levels—use gas separator if above perforations
- Excessive silt, sand, and coal fines causing fluctuations in well inflow
- The well is being pumped off, and the unit must be slowed before gas through pump causes high temperature damage to the stator
- Overpressuring has damaged the stator elastomer

13.6 SUMMARY

- PCPs have advantages, such as handling solids and viscous fluids, high power efficiency,[7] and a relatively low surface profile.
- They are well suited for de-watering coal seam wells.
- They can be used to deliquefy gas wells; however, care must be taken not to pump the fluid level to the pump and have the pump produce with gas, even for a short time.

REFERENCES

1. ABB Presentation on World Wide Artificial Lift Statistics by ABB Automation Technology Products, December 14, 2001.

2. Klein, S. T., "Selecting a Progressive Cavity Pumping System," Southwestern Petroleum Short Course, Lubbock, TX, April 20–21, 1994.

3. Adair, R. L., and Kramer, T., "New Technologies for Progressive Cavity Pumps," Southwestern Petroleum Short Course, Lubbock, TX, April 25–26, 2001.

4. McCoy, J. N., "Analysis and Optimization of Progressing Cavity Pumping Systems by Total Well Management," paper presented at the 2nd SPE Progressing Cavity Pump Workshop, Tulsa, OK, November 19, 1996.

5. Griffin Pumps Operators Manual, 5654 55th Street, Calgary, S. E., Alberta, Canada, T2C 3G9.

6. Weatherford ALS Progressive Cavity Pump Manual, 2001.

7. Saveth, K. J., "Field Study of Efficiencies Between Progressing Cavity, Reciprocating and Electric Submersible Pumps," SPE 25448 presented at the Production Operations Symposium, Oklahoma City, OK, March 21–23, 1993.

CHAPTER 14

OTHER METHODS TO ATTACK LIQUID-LOADING PROBLEMS

14.1 INTRODUCTION

Several methods for removing water from gas wells have been presented, and the methods have been classified under several broad categories. However, there are several existing methods and some new developments that merit discussion but do not fall neatly into any of the previous discussions.

This chapter contains information on:

- Use of ESP cable to add thermal energy to the flowstream
- Casing liners to reduce heat loss from the flowstream
- Thermal coatings to reduce heat loss from the flowstream
- Use of a vacuum in the casing-tubing annulus with a packer installed
- Cycling
- Tubing/annulus flow switching
- Tubing flow controllers
- Tubing inserts to re-integrate liquids into a mist-like flow

14.2 THERMAL METHODS FOR WATER OF CONDENSATION

Chapter 1 discussed several sources of produced water from a gas well. When the water enters the wellbore as water vapor with the gas and condenses further up the wellbore, there are some methods that can be used to keep the water in the vapor phase and prevent condensation in the tubing.

Figure 14-1 shows a gas well with water of condensation. Water enters the wellbore with the gas in the vapor phase and, depending on the

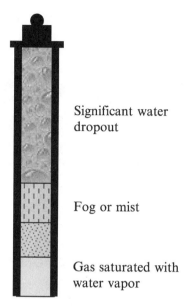

Significant water
dropout

Fog or mist

Gas saturated with
water vapor

Figure 14-1. Gas well with water of condensation.

pressure and temperature variation in the tubing, liquid water may condense out of the vapor phase if wellbore conditions drop below the dew
point for the saturated gas.

In the regions where liquid drops out, the pressure gradient will be
higher than if the water remained in the vapor phase, resulting in a higher
bottomhole pressure (Figure 14-2). If the water could be maintained in
the vapor phase, liquid dropout could be prevented. This would not only
increase production rate by lowering BHP, but also reduce operations
costs and potential corrosion problems.

The temperature profile in a flowing well is determined by several
factors: the reservoir temperature, production rate and thermal properties of the rock, tubulars, and produced fluid. Most of these factors are
beyond the control of the engineer; however, if the temperature of the
gas could be increased sufficiently, either by adding heat energy to the
fluid or by reducing heat loss from the fluid, then the gas stream could
be maintained above the dew point until the gas reaches the surface.

Figure 14-3 shows a well making about 400 Mscf/D gas and 2 bbl/
MMscf water with reservoir temperature = 225°F and flowing wellhead
temperature of 80°F. Water dropout is predicted to occur at about
3500 feet. This depth is indirectly confirmed by observation of tubing
corrosion beginning at this depth.

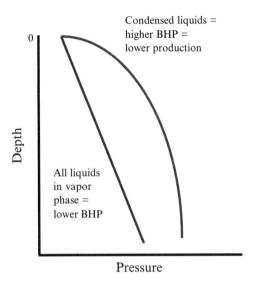

Figure 14-2. Schematic of pressure profile with and without water condensation.

Figure 14-4 shows the same well with a wellhead temperature of 119 °F. At this temperature, water dropout is delayed to just below the surface at 500-ft depth. A wellhead temperature of 120 °F would be sufficient to keep the gas above the dew point in the tubing string, allowing the liquid to condense in the surface flowline or separator.

14.2.1 Thermal Lift

Thermal lift is a method from Centrilift that utilizes modified ESP cable strapped to the outside of the tubing. Reference 1 presents a case study of the application of this method.

Electricity through the cable generates thermal energy to heat the flowing gas above the dew point. This cable arrangement is not thermally efficient without insulation, approximately 80% of the generated thermal energy is lost to the formation through the annulus. A coiled tubing version of the cable is under development that would allow the cable to be run inside the tubing, increasing thermal efficiency and reducing power consumption.

Figure 14-5 illustrates the thermal lift concept. Cable length, size, and required power are calculated for the well design. Because this method can be relatively expensive, overall economics must be carefully evaluated. In the well of Reference 1, operating costs were $5000/month with the heating

Liquid fraction–pressure with gray (Mod)

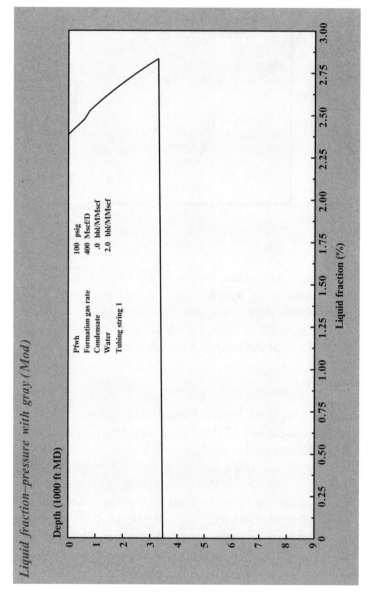

Figure 14-3. Example well with water condensation at 3500 ft with surface temperature of 80°F.

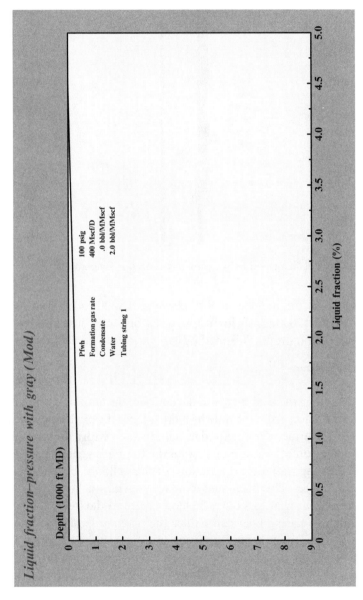

Liquid fraction–pressure with gray (Mod)

Pfwh	100 psig
Formation gas rate	400 Mscf/D
Condensate	.0 bbl/MMscf
Water	2.0 bbl/MMscf
Tubing string 1	

Depth (1000 ft MD)

Liquid fraction (%)

FIGURE 14-4. Example well with surface temperature of 119 °F showing liquid dropout at about 500 ft.

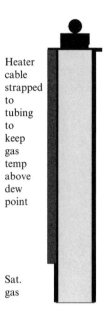

Heater
cable
strapped
to
tubing
to
keep
gas
temp
above
dew
point

Sat.
gas

Figure 14-5. Thermal lift method from Centrilift to keep gas temperature above dew point.

cable strapped to the outside of the tubing and wellhead compression compared with $1200/month for offset wells but were justified by the increased production rate. This method has yet to be widely applied.

14.2.2 Thermal Liner

Instead of adding thermal energy to the fluid stream, another option is to reduce the energy loss from the fluid as it flows from the high-temperature formation to the surface. One method to achieve this is with a thermal liner.

During operation of a flowing gas well, the tubing is generally in contact with the casing wall for large sections of the wellbore, even for nominally vertical wells. This steel-steel contact provides a highly thermal conductive path for heat loss from the flowstream to the lower temperature formation. A casing liner can reduce the magnitude of heat loss by insulating the flowstream from the casing wall.

A thermal liner is a polymer casing insert with insulating properties. The liner in Figure 14-6 is produced by the Polybore Division of Trican Production Services. The liner has an inner layer for corrosion resistance and low permeation and an outer layer for higher temperature strength and insulation. Specialty thermal liner designs have achieved thermal conductivity values on the order of 1% of steel.

Figure 14-6. Thermal liner.

Thermal casing liners are still in the development stage but show promise for gas wells with water of condensation. Additional advantages of the liner, such as corrosion and erosion control and possible down-hole injection path (e.g., for foamers, inhibitors, gas lift), may make the economics more attractive.

Trican (Alberta, Canada) has a system that can line tubing with a wear-resistant liner or a liner to resist heat flow. The liner can be installed from the surface without pulling the tubing. A casing liner is available but without the same voids to obtain better insulation as found in the tubing liner.

14.2.3 Thermal Coatings/Liners

Another method to reduce heat loss from the flowstream is to apply a thermal insulating coating to the outside of the tubing, or a liner to the inside of the tubing.

Tenmar (The Woodlands, TX) markets a thermal coating consisting of multiple layers of thin, insulating films. Each layer consists of an epoxy coating with thickness from 0.01 to 0.02 inches. The limit to the number of layers is determined by the desired insulation. The coating material is compatible with produced fluids up to 300 °F and 5500 psi. According to Tenmar, tests in 3.5-inch tubing inside 9.625-inch casing, with air as the annulus fluid, showed the heat loss with the tubing coating to be 36% of the heat loss of uncoated tubing.

14.2.4 With Packer Installed, Draw a Vacuum on the Annulus

Another method of insulating the tubing or a heat source near the tubing is to put a vacuum pump on the annulus with a packer installed. The vacuum should drastically reduce conduction and convection between the tubing and a possible heat source and the casing. This could drastically reduce the heat transfer to the formation. Although this seems like a good idea for insulating the tubing, one field case reported less insulation effect than predicted.

14.3 CYCLING

When a gas well begins to liquid load, it can often be operated by switching from flow to shut-in conditions or by cycling the well. Cycling may also be called stop-cocking or intermitting. Shutting the well in allows reservoir pressure to build and pressured gas to accumulate in the annulus. After the casing pressure has increased sufficiently, the well is opened to flow. The pressured casing gas expands into the tubing and displaces any remaining accumulated liquids to the surface. During the shut-in period, the well pressure pushes all or part of the fluids back into the formation, allowing gas to flow when the well is opened.

As we have seen in Chapter 7, cycling is very similar to plunger lift, except without a plunger. If a well can be cycled in this fashion, it can also be operated more efficiently on plunger lift (provided there are no physical limitations to prevent plunger operation). Economics will dictate whether or not plunger operation will provide sufficient increased production to justify the additional operating cost. Again, if the well can produce a slug of liquid at or above about 1000 ft/min a plunger may not help. If flow lowers slug velocities, a plunger should improve production.

Some wells can be cycled without production tubing. When flowing up casing only, it may be possible to cycle the well if the reservoir pressure can build to a sufficiently high pressure to unload the well, and the reservoir capacity can produce an initial gas velocity above the critical velocity for the well conditions. The initial high rate will unload the well, allowing the well to continue to flow until the reservoir pressure declines, and the gas rate drops below the critical rate and the well begins to liquid load again.

When cycling a well, the well must not be allowed to flow for too long below the critical rate; otherwise, too much liquid may accumulate in the wellbore, making it difficult or impossible to unload by cycling. If this occurs, the well must be unloaded by other means before cycling can resume.

A well-operated cycling can be controlled by simple timers to shut-in for a set period, then flow for another period.

Cycling can also be controlled with more sophisticated controllers that monitor the casing and tubing pressures and gas rate when flowing, similar to plunger lift controllers.

As the static reservoir continues to decline, cycling will become more difficult. Eventually, cycling will fail, and a decision must be made to apply other methods.

14.4 TUBING/ANNULUS SWITCHING CONTROL

In wells that are tubing restricted, where the flow up the tubing is above the critical rate but also generates excessive friction pressure loss, it may be possible to achieve increased production by flowing up the

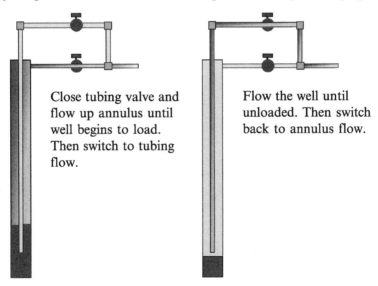

Close tubing valve and flow up annulus until well begins to load. Then switch to tubing flow.

Flow the well until unloaded. Then switch back to annulus flow.

Figure 14-7. Tubing/annulus control.

annulus, where friction is low, until the well begins to load up. When liquid loading is detected, flow is switched back to the tubing to unload the well. Figure 14-7 illustrates this concept.

Cycling back and forth between tubing and annulus flow, without shutting in the well, will allow lower average flowing bottomhole pressure and increased production.

14.5 TUBING FLOW CONTROL

Tubing flow control[2] is similar to tubing/annulus switching, except that flow is continuous and does not switch back and forth between annulus and tubing.

The object of tubing flow control is to control the casing pressure so that just enough gas is allowed to flow up the tubing to maintain the rate above critical while any excess gas is flowed up the annulus. Excess tubing blow-downs or large fluid levels in the casing (if no packer) are prevented, a good rule of thumb[2] to select well candidates for tubing flow control is:

Well Gas Rate $\geq$ 2 $\times$ Well Critical Rate

Figure 14-8 illustrates the tubing flow control concept. Control is achieved by adjusting the casing choke to split the gas flow between

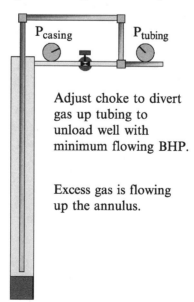

P_{casing} P_{tubing}

Adjust choke to divert
gas up tubing to
unload well with
minimum flowing BHP.

Excess gas is flowing
up the annulus.

Figure 14-8. Tubing flow control.

tubing and annulus so that the gas flow up tubing is at or above the tubing critical rate and flowing bottomhole pressure is minimized. Annulus and tubing flow rates are controlled and optimized but not cycled. The flow is continuous, and the well is never shut-in.

Tubing flow control can be automated to monitor the tubing flow rate, and the casing choke can adjusted automatically to regulate tubing flow at the optimum flow to minimize flowing bottomhole pressure and maximize total production (tubing + annulus).

14.6 TUBING COLLAR INSERTS FOR PRODUCING BELOW CRITICAL VELOCITY

Putra and Christiansen[3] proposed a new method to enhance well performance when flowing below critical velocity in the churn flow regime.

The method uses inserts spaced throughout the tubing string (Figure 14-9). The inserts are specially designed restrictions to induce liquid droplet formation and to prevent liquid fallback. The combination of these two effects redistributes the liquid more evenly along the tubing, reducing overall pressure drop.

The study defines the upper and lower limits on gas rate for application of the inserts, so that the method is effective only when flowing in churn flow within the upper and lower gas rate limits. This method is a new development and has not yet been fully field tested.

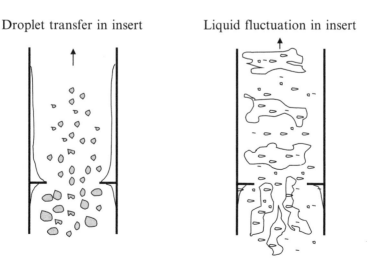

Droplet transfer in insert   Liquid fluctuation in insert

Figure 14-9. Tubing collar inserts.

14.7 SUMMARY

- The effects of water of condensation can be reduced or eliminated by keeping the temperature of the flowstream above the dew point to prevent water dropout in the tubing.
- Use of thermal lift (ESP heating cable) to add thermal energy to the flowstream or use of casing liners or thermal coatings to reduce heat loss from the flowstream can be effective in increasing the flowing temperature.
- If a packer is installed, pulling a vacuum on the annulus should reduce heat transfer from the tubing and a possible heat source near the tubing to the casing and to the formation. No cases are reported, however.
- Cycling a well by periodically shutting in the well to build pressure and then opening the well to flow to unload the well can be effective for wells that cannot flow continuously without liquid loading.
- Switching back and forth from tubing flow to unload the well, then to annulus flow to produce the well can be an effective way to increase production in a well that is tubing friction limited.
- Controlling the annulus pressure to regulate the tubing flow rate to minimize flowing bottomhole pressure and increase production can be an effective method.
- Use of specially designed restriction to induce droplet formation and redistribute liquid throughout the tubing may be effective when producing below critical velocity.

REFERENCES

1. Pigott, M. J., Parker, M. H., Vincente, D., Dalrymple, L. V., Cox, D. C., and Coyle, R. A., "Wellbore Heating to Prevent Liquid Loading," SPE 77649, presented at the SPE Annual Technical Conference and Exhibition, San Antonio, TX, September 2, 2002.

2. Elmer, W. G., "Tubing Flowrate Controller: Maximize Gas Well Production from Start to Finish," SPE 30680, presented at the SPE Annual Technical Conference & Exhibition, Dallas, TX, October 22–25, 1995.

3. Putra, S. A., and Christiansen, R. L., "Design of Tubing Collar Inserts for Producing Gas Wells Below Their Critical Velocity," SPE 71554, presented at the SPE Annual Technical Conference & Exhibition, New Orleans, LA, September 30–October 3, 2001.

DEVELOPMENT OF CRITICAL VELOCITY EQUATIONS

A.1 INTRODUCTION

This Appendix summarizes the development of the Turner[1] equations to calculate the minimum gas velocity to remove liquid droplets from a vertical wellbore.

A.1.1 Physical Model

Consider gas flowing in a vertical wellbore and a liquid droplet transported at a uniform velocity in the gas stream as illustrated in Figure A-1.

The forces acting on the droplet are gravity, pulling the droplet downward, and the upward drag of the gas as it flows around the droplet.

The gravity force is:

$$F_G = \frac{g}{g_C}(\rho_L - \rho_G) \times \frac{\pi d^3}{6}$$

and the upward drag force is given by:

$$F_D = \frac{1}{2g_C}\rho_G C_D A_d (V_G - V_d)^2$$

where   g = gravitational constant = 32.17 ft/s^2
   g_C = 32.17 lbm-ft/lbf-s^2
   d = droplet diameter
   ρ_L = liquid density

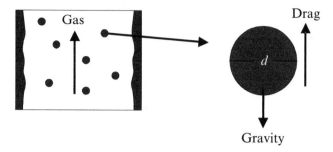

Figure A-1. Liquid droplet transported in a vertical gas stream.

ρ_G = gas density
C_D = drag coefficient
A_d = droplet projected cross-sectional area
V_G = gas velocity
V_d = droplet velocity

The critical gas velocity to remove the liquid droplet from the wellbore is defined as the velocity at which the droplet would be suspended in the gas stream. A lower gas velocity would allow the droplet to fall, resulting in liquid accumulation in the wellbore. A higher gas velocity would carry the droplet upward to the surface and remove the droplet from the wellbore.

Thus, the critical gas velocity V_C is the gas velocity at which $V_d = 0$. In addition, because the droplet velocity is zero, the net force on the droplet also is zero. The defining equation for the critical gas velocity is then:

$$F_G = F_D$$

or:

$$\frac{g}{g_C}(\rho_L - \rho_G)\frac{\pi d^3}{6} = \frac{1}{2g_C}\rho_G C_D A_d V_C^2$$

Substituting $A_d = \pi d^2/4$ and solving for V_C gives:

$$V_C = \sqrt{\frac{4g(\rho_L - \rho_G)d}{3\rho_G C_D}} \tag{A-1}$$

This equation assumes a known droplet digameter. In reality, the droplet diameter is dependent on the gas velocity. For liquid droplets entrained in a gas stream, Reference 2 shows that this dependence can be expressed in terms of the dimensionless Weber number:

$$N_{\text{WE}} = \frac{V_C^2 \rho_G d}{\sigma g_C} = 30$$

Solving for the droplet diameter gives:

$$d = 30 \frac{\sigma g_C}{\rho_G V_C^2}$$

Substituting into Equation A-1 gives:

$$V_C = \sqrt{\frac{4 (\rho_L - \rho_G)}{3} \frac{g}{\rho_G} \frac{g}{C_D} 30 \frac{\sigma g_C}{\rho_G V_C^2}}$$

or

$$V_C = \left(\frac{40 g\, g_C}{C_D} \right)^{1/4} \left(\frac{\rho_L - \rho_G}{\rho_G^2} \sigma \right)^{1/4}$$

Turner assumed a drag coefficient of $C_D = .44$ that is valid for fully turbulent conditions. Substituting the turbulent drag coefficient and values for g and g_C gives:

$$V_C = 17.514 \left(\frac{\rho_L - \rho_G}{\rho_G^2} \sigma \right)^{1/4} \text{ft/s} \tag{A-2}$$

where ρ_L = liquid density, lbm/ft^3
$\qquad \rho_G$ = gas density, lbm/ft^3
$\qquad \rho$ = surface tension, lbf/ft

Equation A-2 can be written for surface tension in dyne/cm units using the conversion $\text{lbf/ft} = .00006852 \text{ dyne/cm}$ to give:

$$V_C = 1.593 \left(\frac{\rho_L - \rho_G}{\rho_G^2} \sigma \right)^{1/4} \text{ft/s} \tag{A-3}$$

where ρ_L = liquid density, lbm/ft^3
ρ_G = gas density, lbm/ft^3
ρ = surface tension, dyne/cm

A.2 EQUATION SIMPLIFICATION

Equation A-3 can be simplified by applying "typical" values for the gas and liquid properties. From the real gas law, the gas density is given by:

$$\rho_G = 2.715 \gamma_G \frac{P}{(460 + T)Z}, \ lbm/ft^3 \qquad (A\text{-}4)$$

Evaluating Equation A-4 for typical values of
Gas gravity γ_G 0.6
Temperature T 120°F
Gas deviation factor Z 0.9
Gives

$$\rho_G = 2.715 \times .6 \frac{P}{(460 + 120) \times .9} = .0031P, \ lbm/ft^3$$

Typical values for density and surface tension are
Water density 67 lbm/ft^3
Condensate density 45 lbm/ft^3
Water surface tension 60 dyne/cm
Condensate surface tension 20 dyne/cm
Introducing these typical values and the simplified gas density Equation A-4 into Equation A-3 yields:

$$V_{C,water} = 1.593 \left(\frac{67 - .0031P}{(.0031P)^2} 60 \right)^{1/4} = 4.434 \frac{(67 - .0031P)^{1/4}}{(.0031P)^{1/2}} \ ft/s$$

$$V_{C,cond} = 1.593 \left(\frac{45 - .0031P}{(.0031P)^2} 20 \right)^{1/4} = 3.369 \frac{(45 - .0031P)^{1/4}}{(.0031P)^{1/2}} \ ft/s$$

A.3 TURNER EQUATIONS

Turner et al.[1] found that for their field data, where wellhead pressures were typically ≥ 1000 psi, a 20% upward adjustment to the theoretical values was required to match the field observations. Applying the 20% adjustment then yields:

$$V_{C,\,water} = 5.321 \, \frac{(67 - .0031P)^{1/4}}{(.0031P)^{1/2}} \text{ ft/s}$$

$$V_{C,\,cond} = 4.043 \, \frac{(45 - .0031P)^{1/4}}{(.0031P)^{1/2}} \text{ ft/s}$$

However, in the original paper by Turner et al.,[1] the coefficients were found to be 5.62 for the critical water velocity equation above and 4.02 for the condensate critical velocity above; however, these values are slightly in error.

A.4 COLEMAN EQUATIONS

Coleman et al.[3] found that Equation A-3 was an equation that would fit their data. This was without the 20% adjustment that Turner et al. made to fit their data at higher average wellhead pressures. So if the "corrected" Turner et al. equations are written without the 20% adjustment, then the Coleman et al. equations can be written as below if the same simplifications and typical data are used as above.

$$V_{C,\,water} = 4.434 \, \frac{(67 - .0031P)^{1/4}}{(.0031P)^{1/2}} \text{ ft/s}$$

$$V_{C,\,cond} = 3.369 \, \frac{(45 - .0031P)^{1/4}}{(.0031P)^{1/2}} \text{ ft/s}$$

REFERENCES

1. Turner, R. G., Hubbard, M. G., and Dukler, A. E., "Analysis and Prediction of Minimum Flow Rate for the Continuous Removal of Liquids from Gas Wells," *Journal of Petroleum Technology*, November, 1969, pp. 1475–1482.

2. Hinze, J. O., "Fundamentals of the Hydrodynamic Mechanism of Splitting in Dispersion Processes," *AICHE Journal*, September, 1955, 1, no. 3, p. 289.

3. Coleman, S. B., Clay, H. B., McCurdy, D. G., and Norris, H. L., III, "A New Look at Predicting Gas-Well Load Up," *Journal of Petroleum Technology*, March, 1991, pp. 329–333.

DEVELOPMENT OF PLUNGER LIFT EQUATIONS

B.1 INTRODUCTION

This Appendix summarizes the plunger lift equations developed in Reference 1 for a plunger lifted well as illustrated in Figure B-1.

B.2 MINIMUM CASING PRESSURE

The minimum casing pressure at the moment that the plunger and liquid slug arrive at the surface is given by:

$$P_{C,min} = (14.7 + P_P + P_C S_V)(1 + D/K) \tag{B-1}$$

where P_P = pressure required to lift the plunger, psi
P_C = pressure required to lift one barrel of liquid and overcome friction, psi
S_V = liquid slug volume above plunger, bbl
K = factor to account for gas friction below the plunger
D = plunger depth, ft

Approximate values for K and P_C are given in Table B-1.
In Equation B-1, K is calculated from:

$$K = \frac{Z(T_{avg} + 460)(OD_{TBG}/12)(2 \times 32.2 \times 144 \times 3600)}{(144/53.3 \times \gamma_G \times f_{gas} \times V^2)}$$

where T_{avg} = the average temperature, °F
Z = the average gas deviation factor

Pressures before release in casing and at arrival at surface

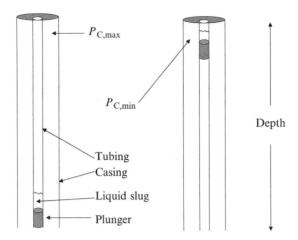

FIGURE B-1. Schematic of plunger lift before release and just as plunger and liquid reach the surface.

Table B-1
Approximate Values for *K* and *P*$_C$ from
Reference 1

Tubing Size (inch)	*K*	*P*$_c$
2.375	33,500	165
2.875	45,000	102
3.000	57,600	67

γ_G = the produced gas gravity
f_{gas} = friction factor for gas flow in tubing
V = average gas velocity, ft/sec
OD_{TBG} = tubing OD, inches

The factor P_C is calculated from:

$$P_C = P_{WEIGHT} + P_{FRICTION}$$

$$P_{WEIGHT} = .433 + \gamma_L \times L_S$$

$$P_{FRICTION} = \frac{62.4\gamma_L f_L L_S V^2}{\{(ID_{TBG}/12) \times 2 \times 32.2 \times 144 \times 3600\}}$$

where $\quad \gamma_L$ = produced liquid gravity
$\quad\quad\quad\quad f_L$ = friction factor for liquid flow in tubing
$\quad\quad\quad\quad L_S$ = liquid slug length
$\quad\quad\quad\quad V$ = average liquid velocity, ft/sec
$\quad\quad ID_{TBG}$ = tubing *ID*, inches

The approach for the development of the minimum casing pressure is that when the slug of liquid and the plunger arrive at the surface of the tubing, the casing pressure must support the weight of the liquid and the plunger, the friction in the tubing, the friction between the liquid, and the tubing and the surface tubing pressure.

B.3 MAXIMUM CASING PRESSURE

The maximum casing pressure is given by:

$$P_{C,max} = \frac{A_{ANN} + A_{TBG}}{A_{ANN}} P_{C,min}$$

where A_{ANN} = cross-sectional flow area between casing and tubing
$\quad\quad\quad A_{TBG}$ = cross-sectional flow area of tubing

This approach assumes conservatively that all energy comes from expansion of the gas from the casing into the tubing to surface the plunger. It can be corrected to account for the gas that is produced as the plunger is coming up to the surface but is not here. It just assumes that when the gas in the casing expands into the tubing, then the surface casing pressure drops to $P_{C,min}$.

B.4 SUMMARY

This development shows what the casing pressure must be at the top of the casing before the tubing valve is opened. The $P_{C,min}$ is the casing pressure when the liquid and plunger arrive. These equations can be used for an oil well. For a gas well, the plunger is then held at the surface to produce gas for some time until the velocity declines to nearly a critical velocity or pressures on the tubing or casing are monitored to reach certain cycle values.

REFERENCE

1. Foss, D. L., and Gaul, R. B., "Plunger Lift Performance Criteria with Operating Experience-Ventura Field," Drilling and Production Practice, API, 1965, pp. 124–140.

A P P E N D I X   C

GAS FUNDAMENTALS

C.1 INTRODUCTION

This Appendix catalogs some commofnly used gas fundamental expressions that are useful when operating gas wells.

C.2 PHASE DIAGRAM

A hydrocarbon gas is a mixture of different hydrocarbon molecules in varying composition. The type and amount of each molecular species in the gas determines the mixture properties at a given pressure and temperature.

As shown in Figure C-1, Critical temperature (T_C) is the temperature of a gas above which it cannot be liquefied by increasing pressure.

Critical pressure (P_C) is the pressure a gas exerts when in equilibrium with the liquid phase at the critical temperature.

Critical volume (V_C) is the volume of one pound of gas at the critical temperature and pressure.

Cricondenbar is the highest pressure at which a gas can exist.

Cricondenterm is the highest temperature at which a liquid can exist.

Bubble point is the pressure at a given temperature above which the mixture is 100% liquid. Dew point is the pressure, at a given temperature, above which the mixture is 100% gas.

C.3 GAS APPARENT MOLECULAR WEIGHT AND SPECIFIC GRAVITY

Molecular weight is defined for a specific molecule but not for a mixture of different molecular species. For gas mixtures, the apparent gas molecular weight M is defined to represent the average molecular weight

Phase diagram

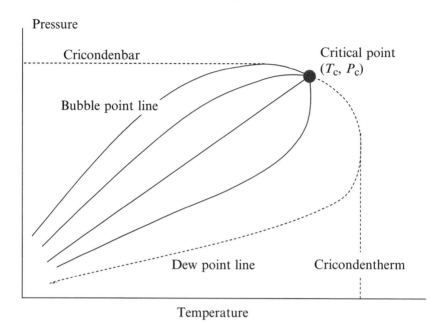

FIGURE C-1. Typical gas well reservoir phase diagram.

of all the molecules in the gas. Thus, M can be calculated from the mole fraction of each molecular species in the gas as:

$$M = \sum_{all\ species\ j} y_j M_j$$

where y_j = mole fraction of molecule j
M_j = molecular weight of molecule j

Example C-1: Molecular Weight of Air

Dry air consists mainly of N_2 (78%, $M = 29.01$), O_2 (21%, $M = 32.00$), argon (1%, $M = 39.94$), and minute amounts of other gases.
Estimate the apparent molecular weight of dry air.

Solution:

$$M_{\text{air}} = \sum_{\text{all species}\,i} y_i M_i = .78 \times 28.01 + .21 \times 32.00 + .01 \times 39.94 = 28.97$$

The specific gravity γ_G of a gas is the ratio of the gas apparent molecular weight to the apparent molecular weight of air.

$$\gamma_G = \frac{M_G}{M_{\text{air}}} = \frac{M_G}{28.97}$$

C.4 GAS LAW

The relationship among pressure, temperature, volume, and density of a real gas is well known and can be described by the gas law equation:

$$pV = nZRT$$

$$\rho = \frac{pM}{ZRT}$$

where p = absolute pressure
$\quad\quad V$ = volume
$\quad\quad T$ = absolute temperature
$\quad\quad n$ = number of moles of gas
$\quad\quad R$ = gas constant
$\quad\quad M$ = molecular weight
$\quad\quad Z$ = gas deviation factor

The gas constant R depends on the units used for the equation as shown in Table C-1.

Table C-1
Gas Constant Values

Units	R
atm-cc/g-mole-°K	82.06
BTU/lb-mole-°R	1.987
psia-ft^3/lb-mole-°R	10.73
lbf/ft^2 abs- ft^3/lb-mole-°R	1544
atm-ft^3/lb-mole-°R	0.73
mm Hg-liters/gm-mole-°K	62.37
in. Hg-ft^3/lb-mole-°R	21.85
cal/g-mole-°K	1.987
kPa-m^3/kg-mole-°K	8.314
J/kg-mole-°K	8414

Example C-2: Density of Dry Air

Estimate the density of dry air at standard conditions (1 atm, 60F).
Solution:
At standard conditions, air is very nearly an ideal gas with $Z = 1$. Then:

$$\rho = \frac{pM}{ZRT} = \frac{14.65 \times 28.97}{10.73 \times (60 + 460)} = .0761 \ \text{lbm/ft}^3$$

C.5 Z FACTOR

An ideal gas would have $Z = 1$. The Z factor, or gas deviation factor, accounts for the deviation of a real gas from ideal gas behavior. The Z factor usually is calculated from correlations based on the gas gravity.

For gas mixtures of chemically similar molecules, the Z factor is correlated with the pseudo-critical temperature T_{pc} and pseudo-critical pressure P_{pc} instead of the actual critical properties.

$$T_{pc} = \sum y_j Tc_j \quad P_{pc} = \sum y_j Pc_j$$

where y_j = mole fraction of gas j.

Note that the pseudo-critical properties are not related to the actual critical temperature and pressure of the gas.

For hydrocarbon gases, the pseudo-critical properties are correlated with the gas gravity as:

$$T_{pc} = 170.5 + 307.3\gamma_G$$

$$P_{pc} = 709.6 - 58.7\gamma_G$$

For condensate fluids:

$$T_{pc} = 187 + 330\gamma_G - 71.5\gamma_G^2$$

$$P_{pc} = 706 - 51.7\gamma_G - 11.1\gamma_G^2$$

From the Law of Corresponding States, all gases have the same Z factor at the same values of reduced temperature T_r and reduced pressure P_r.

$$T_r = \frac{T}{T_c} = P_r = \frac{P}{P_c}$$

Using this concept, a chart for the Z factor of gas mixtures has been developed by Standing and Katz[1] to give the Z factor for values of T_r and P_r. Several equations have been fitted to this chart to explicitly calculate the Z factor.

One equation from Brill and Beggs[2] and corrected by Standing[3] is

$$Z = A + (1 - A)exp^{(-B)} + CP_r^D$$

where $A = 1.39(T_r - .92)^{.5} - .36T_r - 0.101$
$B = P_r(.62 - .23T_r) + P_r^2[0.066/\{T_r - 0.86\} - 0.037] + .32P_r^6/\{exp\,[20.723(T_r-1)]\}$
$C = 0.132 - 0.32log(T_r)$
$D = exp(0.715 - 1.128T_r + 0.42T_r^2)$

Using the above equations, the chart of Figure C-2 was constructed with some sample relationships for Z vs. T_r and P_r.

For impurities, corrections can be made to P_{pc} and T_{pc} according to work done by Wichert and Aziz.[4]

$$T'_{pc} = T_{pc} - \epsilon$$

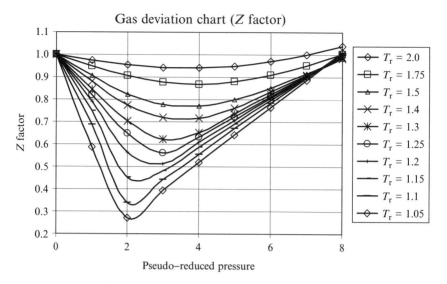

FIGURE C-2. Gas Z factor as function of T_r and P_r.

$$P'_{pc} = \frac{P_{pc} T'_{pc}}{T'_{pc} + \epsilon(B - B^2)}$$

$$\epsilon = 120(A^{0.9} - A^{1.6}) + (B^{0.5} + B^4)$$

where B = mol fraction of H_2S
A = mol fraction $CO_2 + B$

Once these corrected values of T_{pc} and P_{pc} are calculated, then the pseudo-reduced values of T_r and P_r can be used in the preceding equations or in charts using T_r and P_r to find the value of Z.

C.6 GAS FORMATION VOLUME FACTOR

The gas formation volume factor B_g is the ratio of the volume of gas at reservoir conditions to the volume of the same mass of gas at standard conditions. Using the gas law presented previously, the gas formation volume factor becomes:

$$B_{\mathrm{g}} = \frac{0.0283zT}{p}\,\mathrm{ft}^3/\mathrm{scf} = \frac{5.04zT}{p}\,\mathrm{bbl/Mscf}$$

where p = pressure, psia
$\quad\;\; T$ = temperature, degrees Rankine

C.7 PRESSURE INCREASE IN STATIC COLUMN OF GAS

Consider an incremental vertical distance (dh) in a static column of gas. Integrating the expression for dP that occurs over distance dh gives an equation for the pressure at depth:

$$dP = \rho \frac{g}{g_c} dh$$

$$dP = \frac{2.7 p \gamma_G \dfrac{g}{g_c}}{144(T + 460)Z} dh$$

$$\int_{P_{\mathrm{top}}}^{P_{\mathrm{bot}}} \frac{dP}{P} = \frac{2.7 \gamma_g \dfrac{g}{g_c}}{144(T + 460)Z} \int_0^H dh$$

$$\ln(P_{\mathrm{bot}}/P_{\mathrm{top}}) = \frac{0.01875 \gamma_g \dfrac{g}{g_c} H}{(T + 460)Z}$$

$$P_{\mathrm{bot}} = P_{\mathrm{top}} \times exp\left(\frac{0.01875 \gamma_g \frac{g}{g_c} H}{(T + 460)Z}\right)$$

The above equation for P_{bot} can be used to calculate the pressure increase down an annulus of a gas-lifted or flowing multiphase flow well or to the fluid level in the annulus for a pumping well. It is more accurate if the calculations are broken up into increments and the temperature and Z factor are the averages for each segment of calculation.

C.8 CALCULATE THE PRESSURE DROP IN FLOWING DRY GAS WELL: CULLENDER AND SMITH METHOD[5]

$$\frac{dp}{dl} = \left(\frac{dp}{dl}\right)_{el} + \left(\frac{dp}{dl}\right)_{f} + \left(\frac{dp}{dl}\right)_{acc}$$

or:

$$\left(\frac{dp}{dl}\right) = \frac{g}{g_c}\rho\cos(\theta) + \frac{f\rho v^2}{2g_c d} + \frac{\rho v dv}{g_c dl}$$

where: θ is the angle from vertical

Ignoring the acceleration term and substituting in the real gas law gives:

$$\left(\frac{dp}{dl}\right) = \frac{pM}{aRT}\left[\cos(\theta) + \frac{f\rho v^2}{2g_c d}\right]$$

$$v = \frac{q}{A} \qquad q = q_{sc}\frac{p_{sc}Tz}{T_{sc}pz_{sc}}$$

Substituting gives:

$$\frac{dp}{dl} = \left[\frac{pM\cos(\theta)}{zRT}\right] + \left[\frac{MTzp_{sc}fq_{sc}}{RpT_{sc}^2 2g_c dA^2}\right]$$

or:

$$\frac{pdp}{zRdl} = \frac{M}{R}\left[\left(\frac{p}{zT}\right)^2\cos(\theta)\right] + C$$

where:

$$C = \frac{8p_{sc}^2 q_{sc}^2 f}{T_{sc}^2 g_c \pi^2 d^5}$$

Separating variables gives:

$$\int_{P_{tf}}^{P_{wf}} \frac{\frac{p}{Tz}\,dp}{\left(\frac{p}{zT}\right)^2 \cos(\theta) + C} = \frac{M}{R}\int_0^{MD} dl$$

Noting $\cos(\theta) = \dfrac{TVD}{MD}$

$$\int_{P_{tf}}^{P_{wf}} \frac{\frac{p}{Tz}\,dp}{\frac{0.001\,TVD}{MD}\left(\frac{p}{zT}\right)^2 \cos(\theta) + F^2} = 18.75\gamma_g(MD)$$

where $F^2 = \dfrac{0.667 f q_{sc}^2}{d}$

Writing the above equation with grouped terms and breaking the well into only two increments for illustration of length $MD/2$ gives:

Upper half of well:

$$18.75\,\gamma_g\,(MD) = (p_{wf} - p_{if})(K_{mf} + K_{if})$$

Lower half of well:

$$18.75\gamma_g\,(MD) = (p_w - p_{mf})(K_{wf} + K_{mf})$$

where p_{wf} = flowing bhp to be solved for
$\qquad p_{if}$ = flowing tubing pressure, input
$\qquad p_{mf}$ = flowing pressure midway in well

$$K = \frac{\frac{p}{Tz}}{\frac{0.001(TVD)}{MD}\left(\frac{p}{Tz}\right)^2 + F^2}$$

The solution can proceed by first calculating Nre, a friction factor f, and p_{mf} by assuming p_{mf} and solving for p_{mf} using the following equation:

$$18.75\gamma_g\,(MD) = (p_{mf} - p_{if})(K_{mf} + K_{if})$$

Since K_{mf} is a function of p_{mf}, the solution is iterative. Once the intermediate pressure is solved for, then P_{wf} can be solved for in the two-segment example. In a real case for accuracy, the solution would be broken into several increments.

C.9 PRESSURE DROP IN A GAS WELL PRODUCING LIQUIDS

One of many correlations for gas wells producing some liquids is the Gray[6] correlation that was developed by H. E. Gray, an employee of Shell Oil Co. API14BM provides insight to Gray's work. It is a vertical flow correlation for gas wells to determine pressure changes with depth and the bottomhole pressure. The method developed by Gray accounts for entrained fluids, temperature gradient, fluid acceleration, and non-hydrocarbon gas components. Well test data are required to make the necessary calculations. As per Gray, for two-phase pressure drop can be defined from the following equation.

$$dp = \frac{g}{g_c}[\xi\rho_g + (1-\xi)\rho_l]dh + \frac{f_t G^2}{2g_c D_{\rho mf}}dh - \frac{G^2}{g_c}d\left(\frac{1}{\rho_{mi}}\right)$$

where ξ = the in-situ gas volume fraction
D = conduit traverse dimension
G = mass velocity
ρ = density
h = depth
p = pressure
g_c = dimensionless constant
f_t = irreversible energy loss

Further, as given in API14BM, ξ can be defined as:

$$\xi = \frac{1 - exp\left\{-2.314\left[N_v\left(1 + \frac{205.0}{N_D}\right)\right]^B\right\}}{R+1}$$

$$B = .0814\left[1 - .0554\, Ln\left(1 + \frac{730R}{R+1}\right)\right]$$

$$N_v = \frac{\rho_m^2 V_{sm}^4}{g\tau(\rho l - \rho g)}$$

$$N_D = \frac{g(\rho l - \rho g)D^2}{\tau}$$

$$R = \frac{V_{so} + V_{sw}}{V_{sg}}$$

where N_v, N_D, and R are velocity, diameter and superficial liquid to gas ratio parameters, which mainly influence the hold-up for condensate wells. In Gray's method, superficial liquid and gas densities are used and a superficial mixture velocity (V_{sm}) is calculated.

The values of the superficial velocities are determined from:

$$V_s = Q/A$$

The Q values for oil and water are from input of bbls/MMscf for the water and the condensate (oil). The conventional liquid holdup H_l, is found as:

$$H_l = 1 - \xi$$

The final Gray equation is untested by the author for:

1. Flow velocities > 50 ft/sec
2. Tubing sizes > 3 1/2″, tested only for tubing ID $1.049 - 2.992$ in
3. Condensate ratios of > 150 bbls/MMscf
4. Water ratios > 5 bbls/MMscf

C.9.1 Calculated Result with Dry Gas and Gas with Liquids

The curve in Figure C-3 shows how calculated curves of flowing bhps appear plotted at the bottom of the tubing.

C.10 GAS WELL DELIVERABILITY EXPRESSIONS

C.10.1 Backpressure Equation

Perhaps the most widely used inflow expression for gas wells is the gas backpressure equation[7]:

Calculated FBHPs: Dry gas and gas with liquids

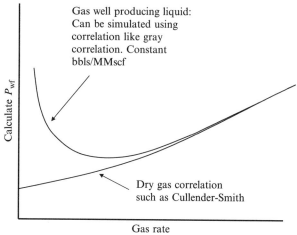

Figure C-3. Tubing performance with/without liquids.

$$q_G = C_1 (\bar{P}_r^2 - P_{wf}^2)^n$$

where q_G = gas rate, units consist with C
C = inflow coefficient
n = inflow exponent
P_r = average reservoir pressure, psia
P_{wf} = flowing bottomhole pressure, psia

Once values for C and n are determined using test data, the backpressure equation can generate a predicted flow rate for any flowing wellbore pressure, P_{wf}. Because there are two constants, C and n, a minimum of two pairs of pseudo-stabilized data (q_g, P_{wf}) are needed but usually at least four data pairs (a "four point" test) are used to determine C and n to account for possible errors in the data collection.
The equation can be written as:

$$\log(\bar{P}_r^2 - P_{wf}^2) = \log \Delta P^2 = \frac{1}{n} \log q_g - \frac{1}{n} \log C$$

A plot of ΔP^2 vs. q_g on log-log paper will result in a straight line having a slope of $1/n$ and an intercept of $q_g = C$ at $\Delta P^2 = 1$. The value of C can also be calculated using any point from the best line through the data as

$$C_1 = \frac{q_g}{(\bar{P}_r^2 - P_{wf}^2)^n}$$

For high permeability wells where the flow rates and pressures attain steady state for each test within a reasonable time (conventional flow-after-flow test), the log-log plot is easily used to generate the needed data. For tighter permeability wells, isochronal[8] or modified isochronal tests and plots can be used where the slope is generated from shorter flow tests, and a parallel line is drawn though an extended pressure-rate point for final results.

To assist in calculating the approximate time for pseudo-stabilized flow to occur starting with any flow, the following equation may be used with $T_{DA} \geq 0.1$. However, for rapidly stabilizing formations, just collect pressure and flow data until they become constant with time.

$$t_s = \frac{380\varphi\bar{\mu}_g\bar{C}_t A T_{DA}}{2.637x10^{-4}k_g}$$

where $\bar{C}_t$ = total compressibility, $1/psi = s_g c_g + s_o c_o + c_f$
with compressibilities and $\bar{\mu}$ evaluated at $\bar{p} = (p_i + p_{wf})/2$.
A = drainage area
φ = porosity

C.10.2 Darcy Equation

Darcy's law for radial flow of a single phase gas is:

$$q_g = \frac{7.03x10^{-4}k_g h(\bar{P}_r^2 - P_{wf}^2)}{\bar{\mu}_g \bar{z}\bar{T}[\ln(x) - 0.75 + s_t + Dq_g]}$$

where q_g = gas flow rate, Mscf/D
P_r = average reservoir pressure, psia
P_{wf} = bottomhole flowing pressure, psia
x = factor related to drainage area geometry (Table C-2)
r_e = radius of external boundary, ft
r_w = radius of the wellbore, ft

Table C-2
Factors for Darcy Equation for Different Shapes and Well Positions in a Drainage Area

System	X	System	X
Circle, well centered	$\dfrac{r_e}{r_w}$	Rectangle (2:1), well centered	$\dfrac{0.966\ A^{1/2}}{r_w}$
Square, well centered	$\dfrac{0.571\ A^{1/2}}{r_w}$	Rectangle (2:1), well off-center	$\dfrac{1.44\ A^{1/2}}{r_w}$
Hexagon, well centered	$\dfrac{0.565\ A^{1/2}}{r_w}$	Rectangle (2:1), well off-center	$\dfrac{2.206\ A^{1/2}}{r_w}$
Triangle, well centered	$\dfrac{0.605\ A^{1/2}}{r_w}$	Rectangle (4:1), well centered	$\dfrac{1.925\ A^{1/2}}{r_w}$
Rhombus 60°, well centered	$\dfrac{0.61\ A^{1/2}}{r_w}$	Rectangle (4:1), well off-center	$\dfrac{6.59\ A^{1/2}}{r_w}$
Right triangle, well at 1/3	$\dfrac{0.678\ A^{1/2}}{r_w}$	Rectangle (4:1), well off-center	$\dfrac{9.36\ A^{1/2}}{r_w}$
Rectangle (2:1), well centered	$\dfrac{0.668\ A^{1/2}}{r_w}$	Rectangle (1:1), well at top	$\dfrac{1.724\ A^{1/2}}{r_w}$
Rectangle (4:1), well centered	$\dfrac{1.368\ A^{1/2}}{r_w}$	Rectangle (2:1), well at top	$\dfrac{1.794\ A^{1/2}}{r_w}$
Rectangle (5:1), well centered	$\dfrac{2.066\ A^{1/2}}{r_w}$	Rectangle (2:1), well off-center	$\dfrac{4.072\ A^{1/2}}{r_w}$
Square (1:1), well at top center	$\dfrac{0.884\ A^{1/2}}{r_w}$	Rectangle (2:1), well off-center	$\dfrac{9.523\ A^{1/2}}{r_w}$
Square (1:1), well off-center	$\dfrac{0.485\ A^{1/2}}{r_w}$	Triangle, two wells	$\dfrac{10.135\ A^{1/2}}{r_w}$

A = Drainage area. $A^{1/2}/r_w$ = Dimensionless.
Courtesy Schlumberger.

k_g = effective permeability to gas, md

$\bar{z}$ = gas deviation factor determined at the average temperature and average pressure

$\bar{T}$ = average reservoir temperature, °R

$\bar{\mu}_g$ = gas viscosity, cp, determined at the average temperature and average pressure

s_t = total skin

Dq_g = pseudo-rate–dependent skin due to turbulence or non-Darcy flow. This is usually zero for low pressure gas wells that might be liquid loaded.

Neely[8,9] rewrote the above single-phase flow equation for gas wells as:

$$\frac{q_G = C(\bar{P}_r^2 - P_{wf}^2)}{\bar{\mu}\bar{z}}$$

where $\bar{\mu}$ = average viscosity that is a function of pressure

$\bar{z}$ = average gas deviation factor that is a function of pressure

C = a constant (not same as C in back pressure equation) and can determined from a single well test if the shut–in average reservoir pressure is known.

The P_{wf} should be determined from a downhole pressure gauge. The viscosity and Z factor should be determined at the bottomhole temperature and average bottomhole pressure. Then C will not change as rates are varied from the well unless damage sets in, such as scale buildup. Using this equation can result in a more accurate inflow expression showing a correction to a higher AOF compared to the old log-log backpressure equation.[11]

REFERENCES

1. Standing, M. B., and Katz, D. L., "Density of Natural Gases," Trans. AIME, 1942.

2. Brill, J. P., and Beggs, H. D., "Two Phase Flow in Pipes," University of Tulsa, Tulsa, OK, 1978.

3. Standing, M. B., "Volumetric and Phase Behavior of Oil Field Hydrocarbon Systems," SPE of AIME, 8th Printing, 1977.

4. Wichert, E., and Aziz, K., "Calculate Z's for Sour Gasses," Hydrocarbon Proceedings, May 1972.

5. Cullender, M. H., and Smith, R. V., "Practical Solution of Gas Flow Equations for Wells and Pipelines with Large Temperature Gradients," Trans. AIME 207, 1956.

6. Gray, H. E., "Vertical Flow Correlation in Gas Wells." User manual for API 14B, Subsurface controlled safety valve sizing computer program, App. B, June 1974.

7. Rawlins, E. L., and Schellhardt, Am. A., "Back Pressure Data on Natural Gas Wells and Their Applications to Production Practices," Bureau of Mines Monograph 7, 1935.

8. Fetkovich, M. J., "The Isochronal Testing of Oil Wells," Paper 4529, 48th Annual Fall Meeting of SPE, Las Vegas, NV, 1973.

9. Neely, A. B., "The Effect of Compressor Installation on Gas Well Performance," HAP Report 65-1, Shell Oil Company, January 1965.

10. Greene, W. R., "Analyzing the Performance of Gas Wells," presented at the annual SWPSC, Lubbock, TX, April 21–22, 1978.

11. Russell, D. G., Goodrich, J. H., Perry, G. E., and Bruskotter, J. F., "Methods for Predicting Gas Well performance," *Journal of Petroleum Technology*, January 1965.

INDEX